NMR

Basic Principles and Progress
Grundlagen und Fortschritte

Volume 8

Editors: P. Diehl   E. Fluck   R. Kosfeld

With 26 Figures

Springer-Verlag

New York · Heidelberg · Berlin 1974

Professor Dr. P. DIEHL

Physikalisches Institut der Universität Basel

Professor Dr. E. FLUCK

Institut für Anorganische Chemie

der Universität Stuttgart

Professor Dr. R. KOSFELD

Institut für Physikalische Chemie

der Rhein.-Westf. Technischen Hochschule Aachen

ISBN 0-387-06618-7 Springer-Verlag New York · Heidelberg · Berlin
ISBN 3-540-06618-7 Springer-Verlag Berlin · Heidelberg · New York

Chemically Induced Dynamic Nuclear and Electron Polarizations-CIDNP and CIDEP

CLAUDE RICHARD

Laboratoire de Chimie Radicalaire
Equipe de Recherche N° 136 associée au C.N.R.S.
Case Officielle N° 140
54037 Nancy Cedex, France

PIERRE GRANGER

Laboratoire de Chimie Théorique
Université de Nancy I
Case Officielle N° 140
54037 Nancy Cedex, France

Contents

Chapter II. The Theory of the CIDNP Effect

Chapter III. Applications to the Study of Chemical Reactions and Magnetic Properties

Chapter IV. The Chemically Induced Dynamic Electron Polarization (CIDEP Effect)

Introduction

Anomalous electron-spin state populations in the Electron Paramagnetic Resonance (EPR) spectra of radicals formed during radiolysis experiments were observed in 1963 by Fessenden and Schuler [170a]. This phenomenon did not receive much attention at the time.

In 1967, Bargon, Fischer, and Johnsen [5] and independently Ward and Lawler [7, 8] reported a similar phenomenon for Nuclear Magnetic Resonance (NMR) spectra taken during radical reactions: emission or enhanced absorption, or both. The earliest attempts to explain this new NMR phenomenon treated these effects in a way similar to that of Dynamic Nuclear Polarization (DNP) or the Overhauser effect. Although the polarization has a completely different origin, DNP gave its name to this effect: Chemically Induced Dynamic Nuclear Polarization (CIDNP). [The name Chemically Induced Dynamic Electron Polarization (CIDEP) was introduced later by analogy with CIDNP]. After the initial publications, all the new data demonstrated that the first theory could not be correct.

In 1969, a new theory was proposed by Closs [18] and independently by Kaptein and Oosterhoff [23] and called the *radical-pair theory*. This mechanism was proposed to account for the observations of polarization in both NMR and EPR.

The radical-pair theory is based on weak interactions in a pair of radicals: the strength of interaction between the electronic states of the radicals depends in particular on the nuclear-spin states.

A great number of experiments were devised in order to test this theory for nuclear polarization, and it is now proved that the radical pair theory can account for the qualitative characteristics of all published CIDNP spectra. This theory has been modified by various authors, but the new developments do not change the basic concepts, and in many cases the theory is capable of quantitatively explaining the data.

CINDP is a powerful tool for physics and chemistry because it gives information about radical reactions and magnetic properties of radicals and molecules. CIDNP can be considered as complementary to EPR, and a valuable addition to the chemist's arsenal.

In contrast, knowledge of CIDEP is not so developed as that of CIDNP. Only a few experiments have been made and a completely satisfactory theory for this phenomenon does not yet exist.

The first chapter is an introduction to the concepts of CIDNP; the basic ideas of the radical-pair theory of CIDNP are presented with the help of an "effective field" model. The second chapter deals with the theoretical developments of the radical-pair theory and is especially interesting to NMR spectroscopists. The third chapter surveys all the chemical applications of CIDNP and is of more particular relevance to chemists interested in radical kinetic problems. The last chapter discusses the CIDEP effect.

Chapter I. Origin of the CIDNP Effect

I. Introduction

A nuclear polarization of the reacting products in the probes of nuclear magnetic resonance (NMR) spectrometers was observed independently in 1967 by two groups of investigators: BARGON, FISCHER, and JOHNSEN (Germany) [5], and WARD and LAWLER (USA) [7, 8]. The NMR spectra of these products show emission or enhanced absorption, disappearing to give lines of normal intensity at the end of the reaction. The nuclear spin states are polarized when they acquire populations different from the Boltzmann equilibrium during this chemical reaction.

Since these reactions involve radicals, this nuclear polarization was thought at the time to originate from electron-nucleus cross-relaxation in the free-radical intermediates. This effect has a close analogy to the dynamic nuclear polarization (DNP) generated in microwave pumping experiments and generally described as the Overhauser effect. The analogy between these two effects gives its name to the new phenomenon: Chemically Induced Dynamic Nuclear Polarization (CIDNP). However, new data later demonstrated that the initally proposed cross-relaxation mechanism was insufficient to explain all results.

In 1969 a new mechanism was proposed by CLOSS [18] and independently by KAPTEIN and OOSTERHOFF [23] and called the radical-pair theory or CKO theory. This theory can explain all published data not only qualitatively, but also quantitatively in many cases. This mechanism is based on the weak interactions between a pair of radicals: with two unpaired electrons, the electronic states of the pair are singlet S (spin angular momentum: 0) or triplet T_σ (spin angular momentum: 1; $\sigma = +1, 0, -1$) states. The transitions between singlet and triplet states occur with nuclear spin-dependent transition probabilities, and give rise to a nuclear spin-state polarization which can be observed by NMR in the products formed according to the following scheme:

$$\boxed{R \cdot \cdot R'} \begin{cases} \nearrow \underline{R} - \underline{R}', \underline{R}(+H) + \underline{R}'(-H) \\ \\ \searrow \underline{R} \cdot + \underline{R}' \cdot \quad \text{subsequent reactions of} \end{cases}$$

polarized radicals (i.e. $\xrightarrow{+SX} \underline{R}X, \underline{R}'X$)

(polarized species are underlined)

Studies of this phenomenon also provide useful information on radical reactions in solution, i.e. the nature and presence of the radical intermediates, the elementary processes giving rise to the polarized species. By traditional kinetic arguments, CIDNP can be considered a reliable tool for the investigation of reaction mechanisms. In the near future, DREIDING predicted [90] that the appropriate question to put to any lecturer would probably be: "Did you observe the CIDNP in your reaction".

From the CIDNP patterns it is possible to derive the magnetic properties of free radicals as signs of hyperfine coupling constant, nuclear relaxation times. Quantitative studies should reveal the magnitudes of these magnetic parameters, and the microscopic behavior of liquids (cage effect, diffusion and so on).

CIDNP effects can be observed with ordinary nuclear magnetic spectrometers. In many cases the reactants are placed in an NMR tube in the spectrometer and the spectrum is drawn as the reaction proceeds. Special applications require only minor modifications.

The first interpretation by cross relaxation process in radicals, in analogy to DNP, gave its name to this phenomenon (The name was first proposed by FISCHER). CIDNP has been retained as a general designation for nuclear polarizations observed during chemical reactions, though the interpretation of the radical-pair theory involves a completely different phenomenon. The observation of nuclear polarization during chemical reactions is often called the CIDNP effect.

Since its discovery in 1967, CIDNP has been the subject of international symposia in 1970 (Houston, Texas) [49], 1971 (Brussels, Belgium) [90, 102], and 1972 (Tallin, Estonian SSR) [212]. Papers presented at the 10th Symposium on Free Radicals in 1971 (Lyons, France) [93] showed it to be a new method for studying radical reactions.

In Europe this phenomenon is often called KIDN(A)P, while it has been referred to as CIDKP (Chemische Induzierte Dynamische Kernpolarisation) in Germany [6] and P(A)NIC (Polarisation Nucléaire (Dynamique) Induite Chimiquement) in France [134, 135].

The CIDNP effect involves the magnetic properties of both electron and nucleus in a magnetic field. These properties concern the basis of EPR (Electron Paramagnetic Resonance) and NMR which are supposed to be known (see for instance [1a, 1d, 1i]).

II. The Phenomenon

We now give two examples of this phenomenon in order to illustrate the two polarization types generally observed in the CIDNP spectra.

1. Examples

BARGON et al. [5] discovered the CIDNP effect during thermal decomposition of peroxides and azo compounds in the preheated probe of an NMR spectrometer. Figure 1.1a shows NMR spectra taken during the thermal decomposition of dibenzoyl peroxide at $110°\,C$ in cyclohexanone: At $t = 0$, the sample has just been transferred into the probe and the NMR spectrum shows the normal absorption lines of dibenzoyl peroxide in the phenyl proton region. During the decomposition reaction these lines vanish and an emission line appears at $\delta = 7.31$ ppm. The amplitude of this line reaches a maximum at $t = 4$ minutes, then decreases. At $t = 7$ minutes, this line reappears in absorption and reaches a constant maximum amplitude at the end of the reaction. Figure 1.1b shows the time dependence of the amplitude of this line.

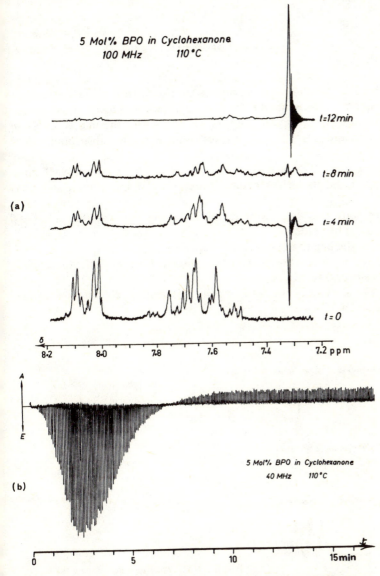

Fig. 1.1. Thermal decomposition of dibenzoyl peroxide a) NMR spectra during reaction b) Time development of the amplitude of the line at $\delta = 7.31 \; 10^{-6}$ (From BARGON et al. [5])

This line has been assigned to the benzene molecules formed through hydrogen abstraction from the solvent SH by transient phenyl radicals:

$$C_6H_5-\underset{\underset{O}{\|}}{C}-O-O-\underset{\underset{O}{\|}}{C}-C_6H_5 \longrightarrow 2C_6H_5-\underset{\underset{O}{\|}}{C}-O\cdot$$

$$C_6H_5-\underset{\underset{O}{\|}}{C}-O\cdot \longrightarrow C_6H_5\cdot + CO_2$$

$$C_6H_5\cdot + SH \longrightarrow \underline{C_6H_6} + S\cdot$$

During this chemical reaction the benzene is formed with a nuclear spin-state population of some protons not in thermal equilibrium, and an *emission* line occurs. At the end of the reaction benzene has relaxed to thermal equilibrium (nuclear relaxation and emission) and it gives rise to a normal absorption NMR spectrum.

WARD and LAWLER have detected nuclear polarization in proton NMR lines from alkenes produced in rapidly reacting mixtures of alkyl lithium and alkyl halides [7, 8, 15].

The spectra shown in Fig. 1.2 were taken during the reaction of *n*-butyl lithium with *n*-butyl bromide:

b) spectrum at the beginning of the reaction;

c) during the reaction, absorption and emission lines appear, approach a maximum intensity and slowly decrease;

d) spectrum at the end of the reaction; this spectrum is consistent with that of 1-butene shown in a).

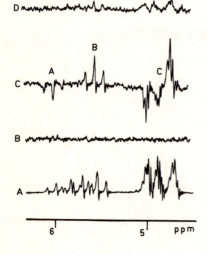

Fig. 1.2. NMR spectra taken during the reaction of *n*-butyl lithium with *n*-butyl bromide (From WARD and LAWLER [8])

Polarized lines observed during this reaction have been assigned to 1-butene formed by the following processes:

$$n\text{-}C_4H_9Br + n\text{-}C_4H_9Li \longrightarrow LiBr + 2CH_3\text{---}(CH_2)_2\text{---}CH_2.$$

$$2CH_3\text{---}(CH_2)_2\text{---}CH_2. \longrightarrow CH_3CH_2\underline{CH}\text{==}\underline{CH}_2 + n\text{-}C_4H_{10}$$

In this case the CIDNP effect is more complex: protons at the 1 position ($\delta = 4.8$) and proton at the 2 position ($\delta = 5.8$) in 1-butene give rise to both *emission and enhanced absorption*.

2. Polarized NMR Spectra

These examples illustrate the two types of CIDNP effect: net polarization, and the multiplet effect: *Net polarization* is observed when equivalent nuclei show NMR lines in *emission* (E) or *enhanced absorption* (A) (first example, net polarization: E). The *multiplet effect* is observed if the normal NMR spectrum of the products shows a multiplet pattern (when the energy levels of the nuclei are altered by interactions with spins of nearby nuclei: spin-spin coupling). The CIDNP multiplet effect is observed when equivalent nuclei show some NMR lines in emission *and* others in enhanced absorption, and for each strong absorption line there is a strong emission line in the multiplet. The pattern of emission followed by enhanced absorption from a lower to a higher magnetic field is called an E/A *multiplet effect*; the opposite case is called A/E. As it is now conventional to present NMR spectra with the field increasing from left to right, for an E/A effect one can observe (from left to right) first emission, then enhanced absorption (second example, multiplet effect: E/A).

The following spectra illustrate these effects:

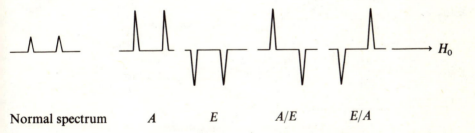

Normal spectrum A E A/E E/A

Net polarization is caused by a deviation of the total Zeeman energy of the spin system from the equilibrium value. Any increase or decrease in this energy gives emission or enhanced absorption. The polarization in this CIDNP effect is an energy polarization. In the multiplet effect, the overall nuclear energy is not far from equilibrium (to a first approximation the integral over the spectrum is zero). This second kind of spin polarization corresponds to a deviation of the entropy of the spin states from the equilibrium value, and can be called an entropy polarization. In fact, as will be seen later, these two types represent the extremes of a broad range of observable CIDNP patterns. If the integral over the spectrum for the transitions originating from the same nuclei is not equivalent to zero[1], the polarization is then positive or negative and is called a Δg *effect*.

III. First Interpretations of the CIDNP Effect

The first theory of CIDNP was the theory of Dynamic Nuclear Polarization (Overhauser effect), where polarization was due to the interaction of the electron with the adjacent protons in the intermediate radical. This theory gave its name to the phenomenon, CIDNP. However it was recognized immediately that this model was insufficient to explain all of the results.

1. Interpretation by Dynamic Nuclear Polarization (DNP)

We shall only give a rapid summary of this interpretation based on the well-known physical effect of DNP [see for instance *1a, 1g, 1h*]. This model was proposed by FISCHER [6, 21] and LAWLER [9].

Figure 1.3 shows an energy-level diagram for a two-spin case consisting of one electron (*e*) and one proton (*n*) in a radical (*HR.*). At thermal equilibrium, Zeeman levels are populated according to the Boltzmann law. These populations are maintained by various relaxation processes with different transition proba-

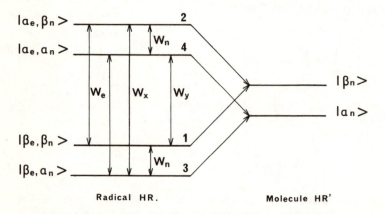

Fig. 1.3. Energy levels of radical (spin system consisting of one electron and one proton) and product molecule (one proton) in a magnetic field

---

[1] More precisely, different from the value given by the thermal equilibrium.

bilities according to the following scheme:

$$(\alpha_e \leftrightarrow \beta_e) \quad W_e e^{\pm \Delta}$$

$$(\beta_n \leftrightarrow \alpha_n) \quad W_n e^{\pm \delta}$$

$$(\alpha_e \beta_n \leftrightarrow \beta_e \alpha_n) W_x e^{\pm(\Delta + \delta)} \quad \Delta = \tfrac{1}{2} \gamma_e \frac{\hbar H_0}{kT}$$

$$(\alpha_e \alpha_n \leftrightarrow \beta_e \beta_n) W_y e^{\pm(\Delta - \delta)} \quad \delta = \tfrac{1}{2} \gamma_n \frac{\hbar H_0}{kT}$$

When free radicals are generated by the rupture of chemical bonds, the unpaired electron spin states have equal populations ($n_1 = n_2$, $n_3 = n_4$). Then the populations of the four Zeeman levels decrease towards a Boltzmann distribution by spin-lattice relaxation. If cross-relaxation transitions are present (W_x and W_y: combined electron spin-nuclear spin relaxation), the nuclear Boltzmann equilibrium may be disturbed, and when the radicals react further the polarization is transferred to the product HR' which gives rise to polarized NMR signals:

$$HR. \xrightarrow{\;W_r\;} \underline{H}R'$$

For observing polarization, the following conditions are required [9]:

$$W_x \text{ or } W_y > W_r > W_e \text{ and } W_n.$$

Polarization depends on the relative values of W_x and W_y. With $W_y > W_x$, States (1) and (2) are overpopulated relative to (3) and (4), and emission (E) results. The predominance of W_y is the result of an anisotropic dipolar coupling in the radical. With $W_x > W_y$, States (3) and (4) are overpopulated relative to (1) and (2) and enhanced absorption (A) results. This situation may arise from a time-dependent scalar coupling between the proton and the unpaired electron.

2. Criticisms of the DNP Theory

The DNP theory was the first attempt to explain CIDNP [6, 9, 21]. It was soon realized that this interpretation could not account for most of the experimental results [16, 17, 67], for example:

(i) the multiplet effect cannot be explained by DNP;

(ii) the magnitude of the observed polarization can be greater than $\gamma_e/\gamma_n \simeq 660$ (γ_e and γ_n being the magnetogyric ratios of the electron and the proton, and γ_e/γ_n the greatest value which can be reached by DNP);

(iii) the electron relaxation rate can be much larger than the rate of reaction of the radical. This leads to the formation of radicals in thermal equilibrium before reaction has taken place;

(iv) products of reaction have polarized protons which do not belong to the free radical:

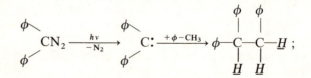

(v) the sign of polarization of a product depends on the formation process of this product (recombination or abstraction reaction).

All these experiments led to the conclusion that nuclear spin polarization occurs in the product-forming steps but not in transient free radicals. This is the radical-pair theory.

3. Other Interpretations

Another attempt to explain the CIDNP effect was made by GERHART and OSTERMANN [136]. In their first work [136a] they explain radical transfer reactions and radical disproportionation reactions, using a transition state with three radicals. For instance, in the reaction:

$$R. + R'X \rightarrow RX + R'.$$

the transition species used was: $[R.X.R'.]$. There is a three-unpaired-electron system. The four energy levels S, T_+, T_0, T_- of the radical pair are split into two levels, giving a total of eight levels. Mixing occurs between symmetrical and antisymmetrical states, giving an increasing population for some levels. From this perturbation of the energy level, qualitative interpretations of the CIDNP effect may be given.

A more quantitative treatment is given by GERHART [136b]. Here, the differences in population between the eight spin-energy levels are supposed to be a function of $\Delta g = g - g'$ (where g and g' refer to radicals $R.$ and $R'.$) and J the electron coupling constant. Several features of the AMX case are given and compared with some experimental results. But no expression is given to correlate measured intensities to physical parameters. Once the CKO theory was developed, this model was no longer used.

IV. The Radical-Pair Theory

We shall now present a qualitative treatment of the radical-pair theory. Theoretical developments of this theory and the modifications by various authors will be treated in the next chapter.

1. Radical-Pair Model

The intermediate of this theory is the radical pair:

$$\boxed{R_1\text{. .}R_2}$$

Let us consider two free radicals formed in close proximity to each other and which coexist in solution for a certain length of time. The inter-radical distance will change because of diffusive displacements and their separation may occur, but the probability of a reassociation is still appreciable, even when the distance attained several molecular diameters. These radicals are said to be the components of a radical pair.

The two radicals can react to give products by combination or disproportionation *(cage products)* or can react with other molecules or radicals when they diffuse in the solution *(escape products)*.

The radical pair can be generated in two ways:

(i) from a common *precursor* molecule in a singlet or a triplet electronic state; this is a *geminate* or *correlated pair*:

$$R_1 - R_2 \rightarrow \boxed{R_1\text{. .}R_2}$$

$$\text{>CN}_2 \xrightarrow{\;-N_2\;} \text{>C:} \xrightarrow{\;+X-C-\;} \boxed{\begin{array}{c} X \\ | \quad | \\ -C.\ .C- \\ | \quad | \end{array}} ;$$

(ii) from a diffusive encounter of two free radicals generated independently:

$$R_1\text{.} + R_2\text{.} \rightarrow \boxed{R_1\text{. .}R_2}$$

These processes are represented in the following scheme:

$$^1M \searrow$$
$$\boxed{R_1..R_2}$$
$$^3M \nearrow$$
$$R_1\cdot + R_2\cdot \Big\}$$

$\nearrow \underline{R}_1 - \underline{R}_2, \underline{R}_1(+H) + \underline{R}_2(-H)$ (cage product)
combination or dismutation

$\rightarrow \underline{R}_1\cdot + \underline{R}_2\cdot$ (escape product)
polarized radicals

This radical pair has two unpaired electrons and two cases are possible for its spin state: triplet state T (spin angular momentum: 1) and singlet state S (spin angular momentum: 0). CIDNP effects can be explained by the two following assumptions:

(i) Transitions between the electronic states S and T with nuclear spin-dependent probabilities are possible during the lifetime of the radical pair. (ii) During radical-radical encounter, reaction occurs only when the radical pair is in a singlet state S giving cage products, this configuration being dependent on nuclear spin. The reaction product acquires a preferred total nuclear spin configuration. The triplet state T does not react and separates with another nuclear spin configuration.

A qualitative picture of this phenomenon can be obtained from the *effective field model* proposed by CLOSS, LAWLER and WARD [*83b, 112, 113*]. In a magnetic field, the threefold degeneracy of the triplet state is lifted and the spin is quantized along the magnetic field direction $(0, \pm 1)$. Figure 1.4 gives a schematic representa-

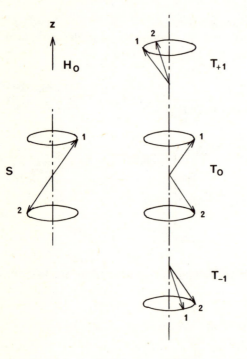

Fig. 1.4. Schematic representation of two weakly coupled electron spins in a magnetic field. The z axis is the direction of the applied field H_0. (From CLOSS, LAWLER and WARD [*83b, 112, 113*])

tion of the motion of the two electron-spin vectors which are weakly coupled to each other. For the singlet state all components of the two-electron spins cancel each other out. We shall only consider transitions between T_0 and S states, which is the most common case in high-field experiments (see Chapter 2).

In a magnetic field H_0, the electron spins precess about the z axis with the Larmor angular frequency ($\omega = g\beta_e\hbar^{-1}H_0$). In the T_0 state the z components of the spins cancel, but there is a resultant spin vector in the x–y plane. For the S state there is a cancellation of all three components. Mixing of S and T_0 is then possible if there is a rotation of one of the electron spin vectors relative to the other about the z axis.

The net magnetic field in the z direction experienced by each electron is the sum of the applied field (H_0) and the internal fields arising from electron orbital motion and nearby nuclear spins. The difference between Larmor frequencies of the two electrons is given in the following equation:

$$\omega_1 - \omega_2 = \beta_e\hbar^{-1}\left[\Delta g H_0 + \sum_i A_{1i}m_{1i} - \sum_j A_{2j}m_{2j}\right].$$

The first term arises from the spin-orbit coupling $\Delta g = g_1 - g_2$, where the g factors of the two electrons are different. The second term arises from electron-nucleus spin coupling (hyperfine splitting in EPR); A and m are hyperfine coupling constants and magnetic quantum numbers of interacting nuclei in radicals $R_1.$ and $R_2..$

In a radical pair, which begins in a triplet state, the projections of the spins in the x–y plane are separated by an angle $\theta = 0$, then a dephasing process takes place (if $\omega_1 \neq \omega_2$). After a certain length of time the projections of the spins cancel out, corresponding to the singlet state (Figure 1.5). This process does not stop here, for after a similar length of time the triplet state is restored. The frequency

T_0 $C_T\,T_0 + C_S\,S$ S

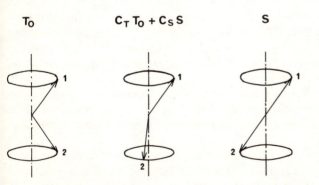

Fig. 1.5. Schematic representation of intersystem crossing from T_0 to S for two weakly coupled electron spins in a magnetic field (From CLOSS [83b])

of this oscillation, or rate of mixing of states depends on the difference Δg, hyperfine coupling constants and nuclear spin states. In typical organic free radicals, g differences and hyperfine coupling constants can give rise to a frequency of dephasing of electron spins (intersystem crossing frequency) of 10^7–10^8 radians per sec. As will be seen in Chapter 2, this value allows mixing of S and T_0 states during the lifetime of a radical pair and can give rise to high magnitudes of polarization.

The effect of this mixing on the polarization of nuclear spin states can be more easily demonstrated by a simplified model system. Let us consider the radical pair $\boxed{HR_1..R_2}$ with $\Delta g = g_1-g_2 > 0$, and one proton in radical 1 coupled to the unpaired electron with a positive hyperfine coupling constant A $(2\beta\hbar^{-1}\Delta g H_0 > A)$:

$$\omega_1 - \omega_2 = \beta_e\hbar^{-1}\Delta g H_0 + mA .$$

If the radical pair is generated in the singlet state, the electron of radical 2 precesses more slowly in the external field H_0 than does that of radical 1. After a certain length of time (which depends on the magnitude of $\omega_1 - \omega_2$), the more rapid precession of the electron of Radical 1 will lead to the T_0 state. However, Radical 1 contains a proton, and intersystem crossing depends on the nuclear states:

$$|\alpha>:m=\tfrac{1}{2} \qquad \omega_1 - \omega_2 = \beta_e\hbar^{-1}\Delta g H_0 + \tfrac{1}{2}A ,$$
$$|\beta>:m=-\tfrac{1}{2} \qquad \omega_1 - \omega_2 = \beta_e\hbar^{-1}\Delta g H_0 - \tfrac{1}{2}A .$$

For nuclear state $|\beta>$ a longer time is required to reach the triplet state. This radical pair with an important singlet character will easily give rise to a cage product. This product will have a more populated $|\beta\rangle$ nuclear state and will show emission (E). The T_0 pairs enriched in $|\alpha\rangle$ nuclear states have a longer life (since reaction is improbable in the triplet state) and consequently have a higher probability of giving diffusing polarized radicals which lead to subsequent reactions (escape products). The escape products are overpopulated in the $|\alpha\rangle$ nuclear state and show enhanced absorption (A). The polarizations of cage products and escape products are of opposite signs.

For T_0–S mixing there is a nuclear spin state which depends on the competition between two pathways of radical destruction (cage or escape products). Only spin selection occurs and no nuclear spin transitions (spin flips) are involved in this polarization process.

A radical pair generated in the triplet state requires a longer time to reach the singlet state for a $|\beta\rangle$ nuclear state. The singlet state will have an overpopulation of the $|\alpha\rangle$ nuclear state and cage products will show enhanced absorption (A). The polarization is the opposite of that resulting from a radical pair generated in a singlet state. Pairs formed by diffusive encounters of free radicals also lead to polarization. The formation of these pairs occurs with random spin distribution, and singlet and triplet states have equal populations. No net intersystem crossing between T_0 and S can be observed in these cases. However, reaction occurs from S states and the remaining pairs will acquire triplet character. In this case the same type of polarization can be expected as from a triplet geminate pair.

As we shall see later, these qualitative ideas point out the two important predictions of the radical pair theory. The polarization sign depends on:

i) radical pair formation; singlet precursor or triplet precursor or pair generated by encounter of free radicals

ii) radical pair destruction: cage product or escape product.

2. Parameters of the Radical-Pair Theory

This qualitative treatment has yielded some parameters on which the CIDNP effect depends. The theoretical developments of Chapter 2 explain the calculation of the polarized spectra and include all the relevant parameters.

KAPTEIN [79] has given rules for the qualitative prediction of polarization in high field experiments (T_0–S mixing) and in two cases: net polarization and multiplet effect. These predictions are based on sign (positive or negative) of the two factors Γ_{np} for a net polarization and Γ_{me} for a multiplet effect, and are given in 2.3.3.b. These rules relate the different parameters on which this effect depends:

(i) the sign of the hyperfine coupling constants;

(ii) the sign of $\Delta g = g_1 - g_2$ (for the net polarization), g_1 and g_2 are usually available from EPR data;

(iii) the mode of formation of the radical pair: singlet or triplet precursor, or encounter of free radicals;

(iv) the mode of destruction of the radical pair: cage product or escape product;

(v) the sign of the nuclear spin-spin coupling constant (for the multiplet effect).

Usually the CIDNP patterns agree with predictions based on these rules. However there are exceptions which are explained by using special computer programs based on the theoretical developments of Chapter 2.

The radical-pair theory involves other parameters.

i) The magnetic field H_0. In high-field experiments there is generally T_0–S mixing, but in low-field experiments $T_{\pm 1}$–S mixing occurs.

ii) The scalar exchange coupling. For a qualitative idea of CIDNP we have assumed that there is no interaction between the electron spins. In fact there is a scalar exchange coupling of magnitude $2J$ which lifts the degeneracy of the S and T_0 states (J is the exchange integral). The time-dependence of J during the lifetime of the radical pair plays an important role in the radical pair theory.

iii) The nuclear relaxation time in the radical. For polarized radicals leaving the radical pair, nuclear relaxation competes with reaction rates with other molecules or radicals.

iv) The nuclear relaxation time in the molecule. If nuclear relaxation is not described by a single relaxation time (T_1) it can affect the CIDNP pattern observed experimentally in the case of a multi-level spin system.

v) The lifetime of the radical pair which involves τ_d, the time of one diffusive displacement ($\simeq 10^{-11}–10^{-12}$ sec).

V. Applications of the CIDNP Effect

Since the CIDNP effect involves interactions in an intermediate radical pair, the transitions between the electronic states depend on nuclear spin states. This effect can give information on the magnetic properties of free radicals and on radical reactions in solution.

1. Determination of the Magnetic Properties of Free Radicals and Molecules

KAPTEIN's rules and qualitative studies of the CIDNP effect give the sign of hyperfine coupling constants, the sign of nuclear spin-spin coupling constants and the relative magnitudes of g factors.

Quantitative studies of this effect can give information on other parameters, e.g. nuclear relaxation times in radicals and molecules, electron-exchange interactions in radical pairs (J), etc. (see Chapter 3).

2. Studies of Radical Reactions

Radical intermediates during reactions in solution can be studied by means of the CIDNP effect. This effect depends on the mode of formation and destruction of the radical pair and can be used for:

(i) determination of spin multiplicity of the precursor: for thermal reactions the difference is generally between geminate and diffusive pairs; for photoreactions, the important differentiation is between singlet and triplet precursors.

(ii) determination of the mode of formation of products: recombination or disproportionation in radical pairs, or subsequent reactions of diffusive polarized radicals leaving the pair.

The CIDNP effect can give information as to the nature of intermediate radicals and radical pairs, etc. (see Chapter 3).

VI. Experimental Procedures

1. Thermal Reactions

This is the usual way in which CIDNP is produced. The probe is preheated to the temperature of the reaction, the sample is introduced rapidly and the spectrum is recorded. This technique is widely used [5, 76].

Difficulties arise when reactions are rapid (several minutes). Spectra cannot be recorded so rapidly. An ingenious modification is proposed by BARGON et al., [5]: a single line is observed and the magnetic field is periodically swept rapidly through the resonance with a sawtooth modulation of 5 to 0.1 Hz. But the recorder cannot move rapidly enough to follow the result of this experiment directly, so the output of the 2 kHz oscillator and the output of the 2 kHz amplifier, which normally go to the lock-in amplifier, are connected to the inputs of a two-channel magnetic tape recorder which records the two output signals simultaneously. The recording speed is 19 cm/sec. When the reaction is finished, the magnetic tape is read at 2.4 cm/sec and fed to the two corresponding inputs of the lock-in amplifier. The signal is then recorded on the spectrometer. Such a result can be seen in Fig. 1.1b. The phenomenon may be displayed on an oscilloscope where it can be photographed [168]. When different species give overlapping multiplet, paramagnetic shift reagents may be used to increase some of the chemical shifts [162].

2. Photochemical Reactions

Generally the probe is not heated. The problem is how to send the light into the sample tube, since the probes are not transparent.

The first idea was to bore a hole into the probe and to irradiate the sample with the focused light of an UV lamp. The first experiments used this technique [13, 16, 48], sometimes with spectrum accumulation [13]. As it is not always possible to modify a probe, other means were proposed [22], i.e. to irradiate the sample outside the magnet and let the liquid flow through a capillary tube located in the coil of the probe. Flow speed was adjusted so that the radicals had not disappeared when they arrived in the coil. This is not a very simple method. MARUYAMA et al. [95] placed their UV lamp above the magnet, and focused the light. This light was guided into the probe through a hole, was reflected once on the aluminium mirror attached within the probe, and arrived at the sample. In order to reduce the loss of light, it was forced to pass through the sample again by means of a concave aluminium mirror. Recently, TOMKIEWICZ and KLEIN [144] used a quartz rod as light guide. The UV light is focused on the top of the NMR tube using a lens and a mirror. A quartz rod with optically polished ends is placed in the tube and enters the solution. This end is placed just above the coil. Solutions which are too concentrated will attenuate the light. This technique does not require modification of the spectrometer and is the best method for photochemical reactions. It has been also applied by ATKINS et al. [214].

3. Other Nuclei

No particular problems occur with other nuclei, except that decoupling techniques [1a, 1b, 1e, 1f] are usually used. In this case the Overhauser enhancement [1g, 1h] occurs and modifies the NMR intensities. This is the case of ^{13}C which will become widely use. As CIDNP is a perturbation of NMR intensities by radicals, such effects as Overhauser enhancement are troublesome and must be suppressed if possible. This can be done by adding paramagnetic molecules to the solution, but in our view, it is inadvisable. A very recent technique which suppresses the Overhauser effect is proposed by FREEMAN et al. [*126*]. Proton decoupling, which is used to observe ^{13}C NMR spectra, is generally continuously applied. These authors, however switch on the proton irradiation just during the 90° excitation pulse for ^{13}C and extinguish it immediately after the beginning of the acquisition of the free induction signal of the ^{13}C for Fourier transformation. This method was used in the study of the reversible addition of pentafluoro-benzoyloxy radical to chlorobenzene [*209*].

Chapter II. The Theory of the CIDNP Effect

I. Basic Principles

Our aim is to interpret the NMR spectra obtained during chemical reactions. We assume that the principles of high-resolution NMR are known to the reader; if not, he may refer to the literature on this subject [1]. We shall mention only the basic facts necessary to understand the theory.

1. Spin Functions

The electron and a number of nuclei possess a magnetic moment which is intimately related to the spin angular moment. This moment is quantized and can take only discrete values. The value of the spin angular moment is characterized by the nuclear spin quantum number I. For the electrons $I = \frac{1}{2}$, for a number of nuclei I may have an integer or half-integer value which is characteristic of the isotope considered. We are only interested in nuclei with I equal to $\frac{1}{2}$ (for example ^1H, ^{13}C, ^{19}F, ^{31}P), and ^2D which has $I = 1$.

Quantum mechanics show that only the magnitude and one component of I can be known simultaneously; generally the z component is chosen. For $I = \frac{1}{2}$, I_z can take two values: $\hbar/2$ and $-\hbar/2$. These two values correspond to two quantum states described by two wave functions α for $I_z = \frac{1}{2}$ and β for $I_z = -\frac{1}{2}$. Each state is characterized by its magnetic quantum number M which takes the value $\pm\frac{1}{2}$ for $I = \frac{1}{2}$.

If several spins interact, the wave function of the system can be constructed using the product of the wave function of each particle. For instance, for three protons $H_1 H_2 H_3$ of a molecule we have $\alpha_{(1)}\beta_{(2)}\alpha_{(3)}$ or $\beta_{(1)}\alpha_{(2)}\alpha_{(3)}$ etc. .... This is generally written as: $\chi = \alpha\beta\alpha$ and $\chi' = \beta\alpha\alpha$.

For each function χ_M which represents a spin state of a system, a magnetic quantum number M is defined. This is the total value of I_z of the system. M is the algebraic sum of I_z values of each individual magnetic quantum number. For instance:

$$\chi_M = \alpha\beta\alpha \rightarrow (\tfrac{1}{2}) + (-\tfrac{1}{2}) + (\tfrac{1}{2}) = \tfrac{1}{2}.$$

$$M = \tfrac{1}{2}$$

It represents also the matrix element:

$$\langle \alpha \beta \alpha | I_{z(1)} + I_{z(2)} + I_{z(3)} | \alpha \beta \alpha \rangle .$$

The case of two electrons is very important and extensively used in the CIDNP theory. For two electrons the basic spin functions are $\alpha\alpha$, $\alpha\beta$, $\beta\alpha$, $\beta\beta$ with $M = 1, 0, 0, -1$, respectively.

We are only concerned here with the spin part of the electron wave function. We may keep in mind that electrons are moving around each nucleus, and their states are described by a molecular-orbital wave function. The two functions $\alpha\beta$ and $\beta\alpha$ have the same M value. They are replaced by two linear combinations: $1/\sqrt{2}(\alpha\beta + \beta\alpha)$ and $1/\sqrt{2}(\alpha\beta - \beta\alpha)$. If we exchange the two electrons, $\alpha\alpha$, $1/\sqrt{2}(\alpha\beta + \beta\alpha)$ and $\beta\beta$ remain unchanged: they are symmetrical and $1/\sqrt{2}(\alpha\beta - \beta\alpha)$ becomes $-1/\sqrt{2}(\alpha\beta - \beta\alpha)$ and is antisymmetrical. These four functions are now divided into two groups. The wave function

$$\varphi_s = 1/\sqrt{2}(\alpha\beta - \beta\alpha) \tag{1a}$$

which describes the singlet state is called S. It represents the normal state of two electrons (Pauli principle). The three other functions represent the three triplet states, T_+, T_0, T_- where:

$$\varphi_{T_+} = \alpha\alpha \quad \text{(1b)} \qquad \varphi_{T_0} = 1/\sqrt{2}(\alpha\beta + \beta\alpha) \quad \text{(1c)} \qquad \varphi_{T_-} = \beta\beta . \quad \text{(1d)}$$

The singlet state has a total spin angular momentum of zero and T states have a total spin angular momentum of 1.

2. Influence of a Magnetic Field on a Particle with a Spin $\frac{1}{2}$

a) Nucleus. A nucleus with spin vector I possesses a magnetic moment $\mu = \gamma\hbar I$ where γ is the gyromagnetic ratio. When it is placed in a magnetic field H_0 which lies along the z axis its magnetic energy becomes:

$$E = -\vec{\mu} \cdot \vec{H}_0 .$$

This energy may be calculated using an Hamiltonian $\mathscr{H}_{Iz}$:

$$\mathscr{H}_{Iz} = -\gamma \hbar I_z H_0$$

where I_z operates only on nuclear-spin functions. E is obtained with:

$$E = \langle \chi_M | \mathscr{H}_{Iz} | \chi_M \rangle = -M \gamma \hbar H_0 .$$

(The action of operators I^2, I_z and the explanation of the notation $\langle | | \rangle$ may be found in textbooks on quantum mechanics [2]). M is the magnetic quantum number previously defined. The most stable state corresponds to $M = \frac{1}{2}$ and hence to an $|\alpha\rangle$ state; $|\beta\rangle$ being the less stable. If the direction of H_0 is reversed, the relative stability of the quantum states may also be reversed.

For a molecule of N nuclei without interactions, the magnetic energy is the sum of the individual energies and is given by:

$$E = -\langle \chi_M | \mathscr{H}_{Iz} | \chi_M \rangle$$

where:

$$\mathscr{H}_{Iz} = -\gamma \hbar H_0 \sum_{i=1}^{N} I_{zi} ,$$

χ_M is the function defined previously.

As each nucleus is surrounded by electrons, the effective magnetic field is not H_0 but H_{eff} which is slightly smaller than H_0. This is called the screening effect.

$$H_{eff} = (1 - \sigma) H_0$$

where σ is the screening constant whose magnitude varies over a range from some 10 to 20 ppm.

Then:

$$\mathscr{H}_{Iz} = -\gamma \hbar H_0 \sum_{i=1}^{N} (1 - \sigma_i) I_{zi} . \tag{2}$$

b) Electron. This is almost the same as for nuclei with spin $\frac{1}{2}$, but the notation is different.

The Hamiltonian is:

$$\mathscr{H}_{Sz} = \beta_e H_0 \sum_{i=1}^{N} g_i S_{zi} \tag{3}$$

where g_i is the spectroscopic splitting factor for the electron i (usually ~ 2 for organic radicals) and S_{zi} is the z component of the spin operator of the ith electron which acts only on electronic spin functions. This operator is equivalent to I_z. (Generally g is a tensor, but in liquids, Brownian motion averages out this tensor.) β_e is the Bohr magneton ($\beta_e = e\hbar/2mc$). The sign of $\mathscr{H}_{Sz}$ is different from $\mathscr{H}_{Iz}$ because electrons have negative charge. The most stable state is $|\beta\rangle$.

In a molecule the basic spin-wave function may be written as the product of an electronic wave function φ_e and of a nuclear wave function χ_M.

The Hamiltonian for the magnetic energy is:

$$\mathscr{H} = \mathscr{H}_{Iz} + \mathscr{H}_{Sz}$$

and the magnetic Zeeman energy for the system is:

$$E_z = \langle \varphi_e \chi_M | \mathscr{H} | \varphi_e \chi_M \rangle.$$

This gives an energy level for each value of the magnetic quantum number of φ_e and χ_M. The splitting is proportional to H_0. Transitions may occur between these levels; their frequencies are given by:

$$h\nu_{ij} = E_i - E_j.$$

Not all transitions are allowed, however. Selection rules show that they are only possible between states of the same symmetry and between energy levels with a total magnetic quantum number whose difference is ± 1. The energy of the transitions is of the order of 10 to 100 MHz for nuclei and 10 to 100 GHz for electrons.

3. Interactions between Spins in a Molecule or a Radical [1a]

For particles with spin $\frac{1}{2}$ there is no quadrupolar moment. Only dipole-dipole interactions are to be considered. They are divided into three groups: nucleus-nucleus, nucleus-electron, electron-electron interactions. We shall consider only the case of liquids.

a) **Nucleus–Nucleus Interactions** (spin-spin coupling J_{ij}). When two dipoles A and B are near one another they interact, and their interaction energy is given by [1f] (p. 15):

$$E = hJ_{AB}I_A \cdot I_B.$$

The Hamiltonian of this interaction is then (in frequency units):

$$\mathscr{H}_{II} = J_{AB}I_A \cdot I_B$$

where $I_A \cdot I_B = I_{xA}I_{xB} + I_{yA}I_{yB} + I_{zA}I_{zB}$.

J_{AB} is the scalar spin-spin coupling constant between nucleus A and B. This coupling is transmitted via bond electrons. Usually $J_{AB} < 500$ Hz and between protons $J_{AB} < 20$ Hz.

For a molecule with several protons:

$$\mathscr{H}_{II} = \sum_{i<j} \sum_j J_{ij}I_i \cdot I_j. \tag{4}$$

b) **Nucleus–Electron Interactions** (A, hyperfine coupling constant). This interaction represents the direct interaction between a nucleus and an electron. This is a magnetic dipole-dipole interaction.

The Hamiltonian of this interaction in frequency units is:

$$\mathscr{H}_{ST} = AS \cdot I \tag{5}$$

where

$$S \cdot I = S_x I_x + S_y I_y + S_z I_z.$$

If the electron or the nucleus is coupled to more than one particle, all the interactions may be added together. A is the hyperfine coupling constant and has a value of several MHz.

c) **Electron–Electron Interaction.** This interaction may be written [1a] (p. 117):

$$\mathscr{H}_{SS} = S_1 \cdot D \cdot S_2 \tag{6}$$

Here D is a tensor which is a function of the positions and the spins of each electron. This interaction removes the degeneracy of $T_\pm$ and T_0 but cannot induce transitions between T and S state. To summarize these different interactions, the energy level diagram of a hydrogen atom placed in a magnetic field is given in Fig. 2.1.

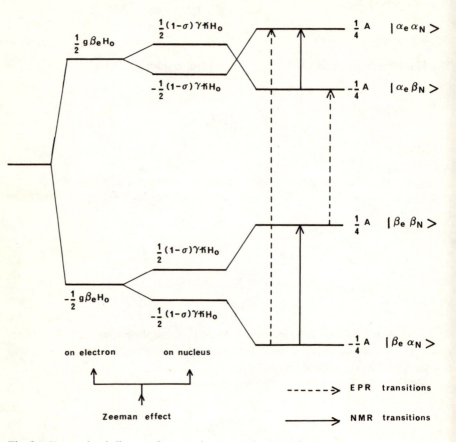

Fig. 2.1. Energy-level diagram for one electron and one nucleus

4. Other Interactions

Magnetic moments are also perturbed by currents. Such currents are produced by electron displacements in the molecules. This explains not only the screening effect and the g factor as already seen, but also other interactions mentioned below.

a) Spin-Orbit Coupling [2a] (p. 118). Electrons in a molecule have an orbital angular momentum which is an eigenvalue of the angular momentum operator L^2. The operator L is related to the rotation of the electrons around the nucleus in an atom where such motion gives rise to a small magnetic dipole which interacts with the different spins. The Hamiltonian of this interaction in frequency units is given by:

$$\mathscr{H}_{LS} = \xi \boldsymbol{L}.\boldsymbol{S} \tag{7}$$

where ξ is the spin-orbit coupling constant. This interaction is small but can mix the ground-state wave function with excited states. $\boldsymbol{L}$ acts on the wave function of the electron and $\boldsymbol{S}$ on the spin function of the electron.

b) Spin-Rotation Coupling. There is a small coupling between the rotation of molecules or radicals with the spins. This interaction is related to the rotational angular momentum $\boldsymbol{J}$ and the different spins of the molecule. The Hamiltonian of this interaction is:

$$\mathscr{H}_{SJ} = \boldsymbol{S}.\boldsymbol{C}.\boldsymbol{J} \tag{8}$$

where $\boldsymbol{C}$ is a tensor. Generally this interaction is small except with unpaired electrons.

5. NMR Line Intensities

As we have already seen in Chapter 1, the CIDNP effect does not disturb the frequency of lines, but their intensities are modified. The theory must explain these variations. It is also important to outline the calculation of NMR line intensities. As seen above, transitions occur only between nuclear spin states described by φ_m and φ_n whose M values differ only by ± 1. These functions φ are eigenfunctions of I_z and I^2 and are linear combinations of the function χ_m previously mentioned. Each function φ_m corresponds to an eigenstate whose energy is E_m (see Fig. 2.2) and the frequency of the transition between the two states is

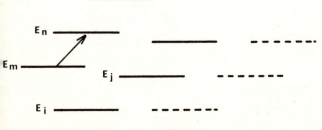

Fig. 2.2. Nuclear energy levels and transition

given by:

$$hv = E_n - E_m.$$

a) Transition Probability. The transition probability is obtained by quantum mechanical calculations [1a, 1b, 1c, 1d, 1e]. The result is:

$$p_{m,n} = |\langle \varphi_n | \sum_{i=1}^{N} I_{xi} | \varphi_m \rangle|^2 \tag{9}$$

where N is the number of nuclear spins and I_x the x component of I.

b) Line Intensities. Experimentally we observe the result of a great number of molecules or radicals. Each energy level E is generally populated according to the Boltzmann distribution. If N_0 is the number of molecules per unit of volume, the number of molecules which have an energy of E_m is:

$$N_m = \frac{N_0 \exp(E_m/kT)}{\sum\limits_{j=1}^{N} \exp(E_j/kT)}$$

where N is the number of energy levels of the spin system.

If the spin system is perturbed, for instance by a chemical reaction, the value of N_m changes to N'_m. The intensity I_{mn} of a line is proportional to the high-frequency power E_{HF} which is absorbed by the spin system

$$I_{mn} = K \left(\frac{dE_{HF}}{dt} \right)_{m \to n}.$$

This energy variation is the product of the energy transition multiplied by the transition probability and the population of the starting level, but we may substract the energy produced by stimulated emission between the same levels. This gives

$$\left(\frac{dE_{HF}}{dt} \right)_{m \to n} = N'_m (E_n - E_m) p_{mn} - N'_n (E_n - E_m) p_{nm}.$$

As $p_{mn} = p_{nm}$ [1a, 1b, 1c, 1d], we have:

$$I_{mn} = K(E_n - E_m) p_{mn} (N'_m - N'_n).$$

As $E_n - E_m$ is approximately the same for all lines of a spectrum (for protons at 60 MHz the differences between frequencies are less than 800 Hz, then $E_n - E_m$ = 60 MHz $\pm$ 400 Hz), the expression of I_{mn} may be simplified:

$$I_{mn} = K' p_{mn}(N'_m - N'_n). \tag{10}$$

The intensity of a line is proportional to the transition probability and to the difference between the population of the two states. If $N_n > N_m$, then $I_{mn} < 0$ and we have an emission line which appears under the baseline of the spectrum obtained experimentally.

6. The Diffusion Theory for Radicals

Apart from quantum interactions in the radical pair, which have been described above, it is necessary to study the movement of radicals in liquids. This leads to several models described in the following sections.

a) The Cage Model. It is assumed that radicals remain close together during a time t. This time is not well defined, but there is a statistical distribution. A function $f(t)$ is used to represent the probability of finding two radicals in contact during a time t.

Generally the following form is proposed:

$$f(t) = e^{-t/\tau_c} \tag{11}$$

which is the classical form for the lifetime of systems such as radioactive nuclei whose evolution is governed by a first-order process. This function was used by CLOSS et al. [18, 41] and KAPTEIN et al. [23]. This model does not agree very well with experimental results.

b) Noyes' Model [4]. Two molecules or radicals which can react, are separated at time $t = 0$. NOYES' theory calculated the probability $f(t)$ for a first encounter of the two particles between time t and $t + dt$. As a general calculation cannot be made, the solution is given for the case where σ, which represents the root-mean-square distance traversed by a molecule during a diffusive displacement, and ϱ, which is the encounter diameter, are equal:

$$f(N) = \frac{0.2390}{(N + 0.442)^{3/2}} \tag{12}$$

N is the number of diffusion steps between $t = 0$ and the time of the first encounter of the two particles. The time between these two events is $t = N/2v$, where v is the frequency with which a molecule makes diffusive displacements, and $2v$ is the frequency of relative displacements between two molecules. For long periods of time, which are of special interest to us, we find that:

$$f(t) = \frac{1.036(1-\beta^2)}{\sqrt{2v}t^{3/2}} \left(\frac{\varrho}{\sigma}\right)^2 \tag{13}$$

with $\beta = \int\limits_0^\infty f(t)dt$, the total probability of at least one re-encounter. An estimation of β is given by NOYES:

$$\beta \simeq 1 - \frac{1}{1/2 + 3/2\,\varrho/\sigma} \tag{13 bis}$$

For small radicals, a single diffusive displacement lasts about 10^{-11} to 10^{-12} second, and $\varrho/\sigma \sim 1$ or slightly > 1. Table 1 of Reference 4 gives β for different values of ϱ/σ. $\beta = 0.527$ for $\varrho/\sigma = 1$, $\beta = 0.815$ for $\varrho/\sigma = 5$ and $\beta \to 1$ if $\varrho/\sigma \to \infty$.

GARST et al. [99] (Footnote 5) proposed the term engagement to describe the set of mutual encounters during the life of the radical pair. The engagement begins with the first collision and its end is determined by the last collision of the two radicals to give reaction products or separate diffusive radicals.

The difference between the two proposed models is in the dependence of $f(t)$ on time. The first is a function of e^{-t/τ_c} and the second a function of $t^{-3/2}$.

II. The Radical-Pair Model

We shall now develop the theory of the CIDNP effect as presented in recent publications [23, 41, 73, 81, 128a, 128b, 176d].

1. Concept of the Radical Pair

The CIDNP effect is observed with the products of radical reactions. The perturbation of the spectra of these products results from the modification of the populations of their energy levels due to the preceding radical reaction. These populations are the result of the evolution of radicals before reaction.

Two radicals, in which the distance between the two parts where the unpaired electrons are located is greater than in a covalent bond but not so much greater as to remove all interactions, form a radical pair.

Examples:

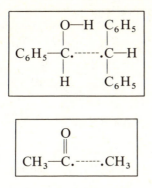

a) **Electronic Quantum States.** Consider a system with two unpaired electrons. From 2.1.1, we have seen that these two electrons may be found in four different states: singlet state S, or triplet states: T_+, T_-, T_0. According to Pauli's principle, a bond may be formed only when the radical pair is in the S state.

b) **Dynamic State.** The interpretation of the movement of the radical in this radical pair is different according to the choice of the model: cage model (2.1.6.a) or NOYES model (2.1.6.b).

In the first case, radicals are supposed to interact inside the cage during time τ_c of Eq. (11). After τ_c the cage is broken and leads to reaction products or free radicals which can react with other radicals in the sample.

In the second case, the radical pair is formed when a molecule is broken into two radicals, or by the meeting of two free radicals. During this short time, interactions occur between the two constitutive radicals. After this initial encounter, the radicals are separated and diffusive jumps begin, without the loss of polarization. There is a new encounter after time t. The probability of an encounter at time t is given by $f(t)$ of Eqs. (12) or (13). During this encounter interactions occur and, if the radical pair is in an S state, the reaction gives a product. If there is no reaction, a new encounter may occur but $f(t)$ decreases rapidly with time and one of the radicals may encounter another one and form another radical pair. This fluctuating system of two radicals forms a radical pair.

2. Radical-Pair Generation

There are two ways to obtain a radical-pair:

a) **From a Common Precursor.** A molecule is cleaved by thermal or photochemical processes, giving two radicals which are close to one another. This forms a radical pair. Thermal dissociations usually give a radical pair in an S state, while photochemical dissociations give singlet or triplet states. As we will see later, this dissociation must be very rapid. For a given dissociation process, the radical pair is always formed in the same state, S or T. We will therefore call the precursor where the radical pair appears in an S state, the S precursor, and the case where it appears in a T state, the T precursor.

b) From Free Radicals. A radical pair may be formed by the encounter of two diffusive free radicals (not polarized). In this case there is the same probability of finding this radical pair in an S state or a T_+, T_0, T_- state. This radical pair is formed from a F precursor.

3. The Disappearance of the Radical Pair

There are two possibilities:

a) Reaction. The radical pair is in an S state when reaction occurs. Recombination products or disproportionation (or dismutation) products are obtained

$$\boxed{R..R} \begin{cases} \nearrow R-R \qquad \text{recombination} \\ \searrow R(+H)+R(-H) \quad \{\text{disproportionation (or dismutation)} \end{cases}$$

In the text these products will be called cage products or c products, terms introduced by FISCHER [76]; KAPTEIN [81, 128a] called them P products.

b) Diffusion. If the electronic state of the radical pair cannot lead to reaction, the two constitutive radicals give rise to two free radicals. These free radicals may encounter another radical and form a radical pair which can lead to products showing the CIDNP effect. These products will be called escape products, or e products, as named by FISCHER [76] (called D products (diffusion products) by KAPTEIN [81]). Figure 2.3 sums up the last three sections devoted to the radical pair.

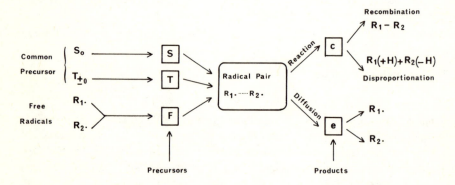

Fig. 2.3. Scheme of radical pair

4. Energy of a Radical Pair in a Magnetic Field

The energy of a radical pair in a magnetic field has been calculated as a function of the distance between the two radicals. Figure 2.4 shows graphically the results of such a computation for S, T_+, T_0, T_- states. A classical case is the study of the H_2 molecule. For an S state we find the classical curve which gives the energy of a bond as a function of the distance between the two atoms. For T states, the magnetic field removes the degeneracy and we obtain three curves corresponding to the three states. Two regions are of particular interest. They are indicated by hatching on Fig. 2.4. In these regions, the energies of the two states are almost equal, or equal, and the system may be considered as partially degenerate. In this case, quantum mechanics shows that the state of the system may be expressed as a combination of the degenerate states. It is then possible to transform a T_- state into an S state in region I, and T_0 into S in region II.

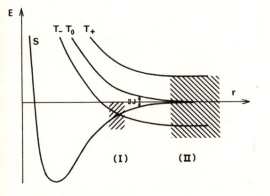

Fig. 2.4. Energy of the different electronic states of the radical pair versus inter-radical distance

As region I is small and the radical pair remains for a very short time at this distance, T–S mixing is not achieved. In region II, which is larger than region I, and where the radical pair remains for a longer time, T_0–S mixing does occur. This mixing between electronic states in region II disturbs the populations of the electronic states and perturbs the nuclear spin states, explaining the variation of intensities in the NMR spectra.

The energy difference between S and T_0 is $2J$, where J is the exchange integral [1a, 1b, 1c]. Values of J are of the same order as, or less than the Zeeman energy or the hyperfine coupling constant A [Eq. (5)].

5. The Hamiltonian of the Radical Pair

a) **The Complete Hamiltonian.** We shall consider a radical pair formed by two radicals a and b. To calculate the total energy of all spins, we must have the Hamiltonian of the system. This question has been studied by ITOH et al. [3]

and their results have been used to interpret the CIDNP effect [*41, 73, 81, 128, 129*]. The total Hamiltonian is:

$$\mathscr{H}_T = \mathscr{H}_e + \mathscr{H}_{LS} + \mathscr{H}_{SZ} + \mathscr{H}_{SS} + \mathscr{H}_{SI} + \mathscr{H}_{Iz} \tag{14}$$

where $\mathscr{H}_e$ is the electronic Coulomb term, and the other terms are those of Section 2.1, i.e. Eqs. (7), (3), (6), (5), and (2), respectively. The most important term is $\mathscr{H}_e$ which can be expressed as a sum of three terms:

$$\mathscr{H}_e = \mathscr{H}_e^{(a)} + \mathscr{H}_e^{(b)} + \mathscr{H}_e^{(ab)}.$$

$\mathscr{H}_e^{(a)}$ is the Hamiltonian of the unpaired electron of radical a, $\mathscr{H}_e^{(a)}$ gives the energy E_0 of this electron and the space wave function ψ_a of this radical. It is the same for $\mathscr{H}_e^{(b)}$. $\mathscr{H}_e^{(ab)}$ is the interaction Hamiltonian between the two radicals. It may be written:

$$\mathscr{H}_e^{(ab)} = \frac{e^2}{r_{12}} - U_{(j)}^{(a)} - U_{(i)}^{(b)} - U^{(ab)}$$

where the two unpaired electrons of the radical pair are labelled 1 and 2; e^2/r_{12} is the electrostatic repulsion energy of these two electrons, $U_{(j)}^{(a)}$ denotes the attraction between the core of radical a and electron j of radical b, $U_{(i)}^{(b)}$ has the same meaning for radical b and electrons i of radical a, and $U^{(ab)}$ is the repulsion between the entire core of radical a and the entire core of radical b. If couplings between electrons and between electrons and orbital angular momentum are neglected, we can use, as basic space wave functions to calculate the energy E_0 from $\mathscr{H}_e$, the two functions:

$$\psi_{Sym} = \frac{1}{\sqrt{2}}(\psi_a + \psi_b) \quad \text{and} \quad \psi_{Anti} = \frac{1}{\sqrt{2}}(\psi_a - \psi_b)$$

where ψ_a and ψ_b are the space wave functions of the unpaired electrons on radicals a and b, respectively. Note that ψ_{Sym} and ψ_{Anti} are space functions but not spin functions. We have:

$$\mathscr{H}_e^{(a)}\psi_a = E_0^{(a)}\psi_a \quad \text{and} \quad \mathscr{H}_e^{(b)}\psi_b = E_0^{(b)}\psi_b.$$

This is now the same problem as for the H_2 molecule [1a, 1c], but in this case, as it is in region II of Fig. 2.4, the distance between the two radicals is large and we may neglect the overlap integral. The interaction energy [1j] (p. 136) is:

$$E_0 = \langle \psi_{Sym} | \mathcal{H}_e^{(ab)} | \psi_{Sym \text{ or Anti}} \rangle = E_0 + K_{ab} \pm J$$

where:

$$E_0 = E_0^{(a)} + E_0^{(b)}, \quad K_{ab} = \langle \psi_a | \mathcal{H}_e^{(ab)} | \psi_a \rangle = \langle \psi_b | \mathcal{H}_e^{(ab)} | \psi_b \rangle$$

and

$$J = \langle \psi_a(1)\psi_b(2) | \mathcal{H}_e^{(ab)} | \psi_a(2)\psi_b(1) \rangle \tag{15}$$

J is a negative integral which is a function of the distance r between the two radicals and is called the exchange integral. In the value of E_0, $+J$ is taken with ψ_{Sym} and $-J$ with ψ_{Anti}. (1) and (2) refer to electron 1 and electron 2.

Now we can take into account the spin of the electron. The new wave function is then the product of a space function ψ with a spin function φ_e. $\Psi = \psi \cdot \varphi_e$. From Pauli's principle this new wave function must be antisymmetrical [1a, 1b, 1c]. Then ψ_{Sym} must be multiplied by an antisymmetrical spin function φ_S which corresponds to the S electronic state of the radical pair. The ψ_{Anti} must be multiplied by a symmetrical spin function which corresponds to the T states. The sign of J is then directly related to the electronic spin function φ_e of the radical pair. The same result is obtained with the following expression of the energy:

$$E_e = E_0 + K_{ab} - \langle \varphi_e | J(\tfrac{1}{2} + 2S_1 \cdot S_2) | \varphi_e \rangle . \tag{16}$$

We can define a new Hamiltonian called the exchange Hamiltonian which is:

$$\mathcal{H}_{ex} = -J(\tfrac{1}{2} + 2S_1 \cdot S_2) . \tag{17}$$

This Hamiltonian has the same form as that for the spin-spin interaction (cf. 2.1.3.c), but is quite different. K_{ab} is spin-independent as $E_0^{(a)}$ and $E_0^{(b)}$. As we are only interested in transitions between spin states, we can choose the zero of the energy, and we shall take $E_0 + K_{ab}$. Then $\mathcal{H}_T$ Eq. (14) becomes:

$$\mathcal{H}'_T = \mathcal{H}_{ex} + \mathcal{H}_{LS} + \mathcal{H}_{SZ} + \mathcal{H}_{SS} + \mathcal{H}_{SI} + \mathcal{H}_{IZ} . \tag{18}$$

Notice that $\mathscr{H}_{ex}$ represents the difference in energy between triplet and singlet states.

b) The Effective Hamiltonian. We are concerned with the study of the value of each energy which appears in Eq. (18) and to leave out the terms which are too small.

i) $\mathscr{H}_{ex}$. This term is very important for short inter-radical distances. It explains the big difference between S and T state energies which appear in Fig. 2.4. If we assume that the time spent outside region II is very short, then we are only concerned with the value of J in region II. It is then assumed that J remains constant. In this case J is of the order of the hyperfine coupling constant A. A detailed discussion is developed by KAPTEIN [81, 128a].

ii) $\mathscr{H}_{LS} + \mathscr{H}_{SZ}$. These two terms are the Zeeman effect and the spin-orbit coupling of each unpaired electron. The Brownian motions average out these interactions and we observe only their mean values. For one electron we may state that:

$$\mathscr{H}_{LS} + \mathscr{H}_{SZ} = \beta_e H_0 g_1 S_{z1}$$

as in Eq. (3). As previously mentioned, this interaction is large and cannot be neglected if the difference between g_1 and g_2 is large. Then for the radical pair:

$$\mathscr{H}_{LS} + \mathscr{H}_{SZ} = \beta_e H_0 (g_1 S_{z1} + g_2 S_{z2}). \tag{19}$$

iii) $\mathscr{H}_{SS}$. We have previously shown (in 2.1.3.c) that this interaction does not mix T and S states, and consequently can be neglected for our purposes.

iv) $\mathscr{H}_{SI}$. This coupling is large and of the same order as J. Its value is given by Eq. (5).

v) $\mathscr{H}_{IZ}$. The nuclear Zeeman effect is very small compared to those of the electrons (see Fig. 2.1). Since σ of Eq. (2) is small and all nuclei have approximately the same value, this effect corresponds to a shift of all energy levels, and can be neglected.

We can write the effective Hamiltonian in frequency units:

$$\mathscr{H} = -J(\tfrac{1}{2} + 2S_1 \cdot S_2) + \beta_e H_0 (g_1 S_{z1} + g_2 S_{z2}) + \sum_i A_{1i} S_1 \cdot I_i + \sum_j A_{2j} S_2 \cdot I_j \tag{20}$$

where i refers to the nucleus of radical a which is associated with electron 1 and j to the nucleus of radical b which is associated with electron 2. This Hamiltonian is used by all authors [23, 41, 73, 81, 128a, 128b]. In [23], A is expressed with the Dirac contact term [1a] which is the most important. KAPTEIN [81, 128a] transforms $\mathscr{H}$ into two subhamiltonians.

6. Spin Evolution of the Radical Pair with Time

Here we are concerned with the study of the changes in the different spins of the radical pair with time. We assume that this pair is in region II of Fig. 2.4. As the major experiments are performed in high magnetic fields, the difference between T_+, T_0, T_- is important and only $T_0 \rightarrow S$ mixing occurs. In the following calculation we will not use T_+ and T_- states.

During the evolution of the radical pair, the inter-radical distance varies and parameters J, g_1, g_2, A_{1i}, A_{2j} change [113, 117]. However, the theory cannot be developed if these parameters are not constant, so we take their mean values. By this procedure J contains the largest error. The other parameters are approximatively constant. The variation of J is outlined by LAWLER [113].

The behavior of the spin system can be obtained from the Schrödinger wave equation [2a, 2b, 2c]:

$$\mathcal{H}\, \psi(t) = -\frac{\hbar}{i}\, \frac{\partial \psi(t)}{\partial t}$$

with $\mathcal{H}$ expressed in frequency units it becomes:

$$\mathcal{H}\, \psi(t) = i\, \frac{\partial \psi(t)}{\partial t}\,. \tag{21}$$

We start at $t = 0$ with an S or a T_0 state and $\mathcal{H}$ mixes these states for $t > 0$. We can use as the spin wave function a linear time-dependent combination of S and T_0 states. We then have:

$$\psi_m(t) = \left(c_{Sm}(t)\varphi_S + c_{T_0m}(t)\,\varphi_{T_0} \right) \chi_m \tag{22}$$

where φ_S and φ_{T_0} are defined in [1a] and [1c] and $\chi_m{}^2$ is the nuclear-spin wave function of the radical pair. We first assume that we have a first-order NMR spectrum [1c, 1f] where χ_m is expressed as a product of the nuclear spin functions. If we incorporate (22) into (21) we obtain two equations whose solutions $c_{Sm}(t)$ and $c_{T_0m}(t)$ are:

$$c_{Sm}(t) = c_S(0)\left[\cos\omega t - \frac{iJ}{\omega}\sin\omega t \right] - ic_{T_0}(0)\,\frac{\mathcal{H}_m}{\omega}\sin\omega t\,, \tag{23}$$

$$c_{T_0m}(t) = c_{T_0}(0)\left[\cos\omega t + \frac{iJ}{\omega}\sin\omega t \right] - ic_S(0)\,\frac{\mathcal{H}_m}{\omega}\sin\omega t \tag{24}$$

---

[2] Note. χ_m and χ_M have different meanings: m is the mth nuclear spin state function of the system; M is the nuclear magnetic quantum number of state χ_m. Several χ_m may have the same M value.

where

$$\omega = \sqrt{J^2 + \mathscr{H}_m^2} \tag{25}$$

$$\mathscr{H}_m = \tfrac{1}{2}\beta_e H_0(g_1 - g_2) + \tfrac{1}{2}\sum_i A_{1i}M_{ai}^{(k)} - \tfrac{1}{2}\sum_j A_{2j}M_{bj}^{(l)} \tag{26}$$

$c_S(0)$ and $c_{T_0}(0)$ are the values of $c_S(t)$ and $c_{T_0}(t)$ at $t = 0$ which is also the moment when the radical pair comes into region II of Fig. 2.4. $M_{ai}^{(k)}$ is the magnetic quantum number of the ith nucleus of radical a in the overall nuclear spin state k. $M_{bj}^{(l)}$ has the same meaning for the jth nucleus of radical b in the overall nuclear spin state l.

We can also obtain $\mathscr{H}_m$ with

$$\mathscr{H}_m = \langle \varphi_S \chi_m | \mathscr{H}' | \varphi_{T_0} \chi_m \rangle$$

and

$$\mathscr{H}' = \tfrac{1}{2}\beta_e(g_1 - g_2)H_0(S_{z1} - S_{z2}) + \tfrac{1}{2}(\mathbf{S}_1 - \mathbf{S}_2) \cdot \left(\sum_i A_{1i}\mathbf{I}_i + \sum_j A_{2j}\mathbf{I}_j \right).$$

The radical pair is found in the S state with a probability given by: $|c_{Sm}(t)|^2$ and in the T_0 state with a probability given by: $|c_{T_0m}(t)|^2$.

Now that we know the behavior of the electron spin states of the radical pair, we may deduce the behavior of the nuclear spin states and explain the CIDNP effect.

First of all it is of interest to give examples of Eqs. (23) and (24). Several such examples may be found in the literature [23, 41, 42, 83b]. We have chosen the example of CLOSS and TRIFUNAC [41].

We consider a radical pair with only one proton which is in radical a. Then χ_m of (22) is one of the two possible states α or β of a proton. There remains only one hyperfine coupling constant between the unpaired electron of the radical a and its proton. If we take $\Delta g = g_1 - g_2$, (26) gives:

$$\mathscr{H}_m = \Delta g/2 \cdot \beta_e H_0 + A/2 \cdot M$$

where $M = \tfrac{1}{2}$ for nuclear spin state α and $-\tfrac{1}{2}$ for β. If we have a T precursor we find it in a T_0 state and then $c_S(0) = 0$ and $c_{T_0}(0) = 1$. If we want $c_S(t)$ we have:

For the nuclear spin state α:

$$c_S^{(\alpha)}(t) = \frac{-i[\tfrac{1}{2}\Delta g \beta_e H_0 + A/4]}{\omega_\alpha} \sin \omega_\alpha t$$

with

$$\omega_\alpha = \sqrt{J^2 + (\tfrac{1}{2}\Delta g \beta_e H_0 + A/4)^2} \, .$$

For the nuclear spin state β:

$$c_S^{(\beta)}(t) = \frac{-i[\tfrac{1}{2}\Delta g \beta_e H_0 - A/4]}{\omega_\beta} \sin \omega_\beta t$$

with

$$\omega_\beta = \sqrt{J^2 + (\tfrac{1}{2}\Delta g \beta_e H_0 - A/4)^2}$$

This gives an oscillating probability. CLOSS et al. [42] have developed the case of two protons, one in radical a and one in radical b.

The exact solution of a two-energy-level system in which the parameter J is modified during the evolution of the radical pair has been computed recently by ATKINS et al. [156, 157] and shows an oscillating probability.

III. Theoretical Calculation of the CIDNP Effect CKO Method
(CLOSS, KAPTEIN, OOSTERHOFF)

1. Reaction Probability

As we know the behavior of the radical pair, we may now calculate the instantaneous probability $P_m(t)$ of reaction of the two components of the radical pair ab with each other, to give the combination or disproportionation products (c products).

All authors [23, 73b, 81, 83b, 128a] consider that the reaction occurs when the radical pair is in an S state, which is the state of the electrons when the bond is formed. $P_m(t)$ is then proportional to $|c_{Sm}(t)|^2$ obtained from (23). This gives the probability of finding a molecule with nuclear spin function χ_m.

Reaction occurs during an encounter of the two radicals of the radical pair, $P_m(t)$ is also a function of the encounter probability $f(t)$, which varies with the choice of the diffusional model.

There remains the proportional factor λ which has been introduced [81, 83b, 128a] and whose value lies between 0 and 1 and represents the efficiency of the encounter. This coefficient may be compared with the primary quantum yield in photochemical reactions. This term will be experimentally determined.

Then:

$$P_m(t) = \lambda f(t) |c_{Sm}(t)|^2 \tag{27}$$

This expression may be calculated with the different models of the radical pair.
a) The Cage Model. When the CKO theory was first presented the cage model was used [18, 23, 41, 103]. It has been developed in 2.1.6.a. We have:

$$f(t) = e^{-t/\tau_c}$$

where τ_c is the lifetime of the cage.

We may consider two cases: S precursor and T precursor.

For an S precursor, the radical pair is in an S state. Then we have $c_{T_0}(0) = 0$ and $c_S(0) = 1$ in (23).
This gives:

$$|c_{Sm}(t)|^2 = 1 - \left(\frac{\mathscr{H}_m}{\omega}\right)^2 \sin^2 \omega t$$

The average value of $P_m(t)$ is:

$$P_m^S = \lambda \left(1 - \frac{2\mathscr{H}_m^2 \tau_c^2}{1 + 4\omega^2 \tau_c^2}\right). \tag{28}$$

For a T precursor with $c_S(0) = 0$ and $c_{T_0}(0) = 1$ we obtain:

$$|c_{Sm}(t)|^2 = \left(\frac{\mathscr{H}_m}{\omega}\right)^2 \sin^2 \omega t$$

and

$$P_m^{T_0} = \lambda \frac{2\mathscr{H}_m^2 \tau_c^2}{1 + 4\omega^2 \tau_c^2}. \tag{29}$$

b) Noyes' Model. This model was first introduced by ADRIAN [73a, 73b, 93c] in the theory of CIDNP, who used Eq. (12) in Section 2.1.6b. It is assumed that the singlet character of a radical pair which separated and returned at the Nth diffusion step is determined by $\sin^2 \mathscr{H}_m \tau_D N$, where τ_D is the mean time between

diffusive displacements, $(\tau_D \simeq 10^{-12}, 10^{-11}$ sec) then the probability of reactions is:
$P_R = \lambda \int_2^\infty |c_{Sm}(\tau_D, N)|^2 f(N) dN$. ADRIAN only considers the case where J is small compared to $\mathcal{H}_m$, so we have $\omega = |\mathcal{H}_m|$.

For a triplet precursor the following value is obtained:

$$P_{Rm}^T = 0.42\,\lambda\sqrt{|\mathcal{H}_m|\tau_D} \tag{30a}$$

and for a singlet precursor:

$$P_{Rm}^S = \lambda(0.31 - 0.42\sqrt{|\mathcal{H}_m|\tau_D}). \tag{30b}$$

CLOSS [83 b] has developed the case where $J = 0$. The expressions are obtained from the preceding one by replacing $\sqrt{\mathcal{H}_m \tau_D}$ with $\dfrac{\mathcal{H}_m^2\sqrt{\tau_D}}{(\mathcal{H}_m^2 + J^2)^{3/4}}$.

If the radical pair is generated by a F precursor, ADRIAN [73b] calculates the recombination probability for the case $J = 0$. Here $c_S(0)$ and $c_{T_0}(0)$ are both different from zero and are complex quantities. The expression of Eq. (23) is:

$$|c_{Sm}(t)|^2 = |c_S(0)|^2 + (|c_{T_0}(0)|^2 - |c_S(0)|^2)\sin^2\omega t$$
$$+ i\,\frac{\mathcal{H}_m}{2\omega}\,(c_{T_0}^*(0)c_S(0) - c_{T_0}(0)c_S^*(0))\sin 2\omega t$$

with $\omega = |\mathcal{H}_m|$ as $J = 0$.

The imaginary part is small and vanishes when the average value is taken for this random phase term. Therefore,

$$P_{Rm}^F = 0.31\,\lambda(\tfrac{1}{2} - \tfrac{3}{8}\lambda) + 0.105\,\lambda\sqrt{|\mathcal{H}_m|\tau_D} \tag{30c}$$

KAPTEIN [81, 128a] has studied this model using Noyes' theory and Eq. (13). The probability of reaction during the first re-encounter at time t is given by $\lambda|c_S(t)|^2 f(t)$, where $f(t)$ is expressed by (13). After this first encounter there may be a second, then a third, and so on. If we use $\lambda_m(t) = \lambda|c_S(t)|^2$, we obtain:

$$P_m(t) = \lambda_m(t)f(t) + \int_0^t \lambda_m(t - t_1)(1 - \lambda_m(t_1))f(t - t_1)f(t_1)dt_1$$
$$+ \int_0^t \Big[\lambda_m(t - t_2)f(t - t_2)\int_0^{t_2}(1 - \lambda_m(t_2 - t_1))(1 - \lambda_m(t_1))f(t_2 - t_1)f(t_1)dt_1\Big]dt_2$$

$$+ \ldots.$$

For an S precursor, λ and $|c_{Sm}(t)|^2$ are close to unity; and only the first term remains in $P_m(t)$. The total probability of reaction is then:

$$P_m^S = \int_0^\infty \lambda |c_{Sm}(t)|^2 f(t) dt$$

which leads to:

$$P_m^S = \lambda [\beta - \mu \pi^{1/2} \mathcal{H}_m^2 \omega^{-3/2}]$$

where

$$\mu = \frac{1.036(1 - \beta^2)}{\sqrt{2v}} \left(\frac{\varrho}{\sigma}\right)^2$$

The parameters are those defined in 2.1.6.b, Eqs. (13) and (13 bis); with

$$x_m = \mu \pi^{1/2} \mathcal{H}_m^2 \omega^{-3/2}$$

we have

$$P_m^S = \lambda(\beta - x_m) . \tag{31a}$$

For a T precursor, $|c_{Sm}(t)|^2$ is small; and the radical pair in state T_0 is slowly transformed into an S state. The only important contribution to $P_m(t)$ comes from terms corresponding to long t.

We then have:

$$P_m^T = \frac{\lambda}{3(1 - \beta)} x_m . \tag{31b}$$

In the case of an F precursor, calculations are more complex because $c_S(0)$ and $c_{T_0}(0)$ are not zero. KAPTEIN [81, 128a] proposes a treatment different from that of ADRIAN [73b].

During the first encounter, we have the probability $\lambda |c_S(0)|^2$ that the pairs combine. Those in a T_0 state may have other encounters, as predicted by Noyes theory. During the following encounters T_0 states may give S states but S states are also transformed into T_0 states. As half of the pairs in the first encounter have $M_S = 0$, the fraction that combines is $1/2\lambda\langle|c_S(0)|^2\rangle = \lambda/4$ where $\langle \rangle$ indicates the mean value. KAPTEIN [81, 128a] (Appendix B) shows that:

$$\langle|c_{Sm}^F(t)|^2\rangle = \tfrac{1}{2}[(1 - \lambda)|c_{Sm}^S(t)|^2 + |c_{Sm}^T(t)|^2] .$$

The recombination probability at the next encounter is given by:

$$\tfrac{1}{2}\lambda \langle |c_{Sm}^F(t)|^2 \rangle f(t).$$

Calculation leads to:

$$P_m^F = \frac{\lambda}{4}\left\{1 + \frac{1}{1 + \beta[1 + \lambda/2(1-\lambda)]}[\beta(1-\lambda) + \lambda x_m]\right\}. \tag{31c}$$

An important case is $\lambda = 1$. Here the expression is simpler:

$$P_{m(\lambda=1)}^F = \frac{1}{4} + \frac{1}{4(1-\beta)}x_m. \tag{31d}$$

The polarization per molecule of product formed is smaller than for a T precursor, but it can be greater than for an S precursor. This is confirmed by experiments [41, 43, 48]. KAPTEIN [81, 128a] shows that there is no CIDNP effect for dimers in thermal equilibrium with their radical monomers.

Recently, BUCHACHENKO et al. [180] have used $f(t) = mt^{-3/2}\exp(-t/\tau)$. This form take into account the fact that the radical pair have a limited lifetime. From the calculation it is shown that for short-living radicals, the nuclear polarization is proportional to $\mathcal{H}_m^2 - \mathcal{H}_n^2$ and for long-living radicals it is proportional to $\mathcal{H}_m^{1/2} - \mathcal{H}_n^{1/2}$ as in Adrians' model.

) **Comparison between the Different Models.** There is a large difference in the equations developed between the cage model (2.3.1.a) and the diffusion model introduced by ADRIAN (2.3.1.b). The first model leads to a function of $\mathcal{H}_m^2$ and the second is a function of $\sqrt{\mathcal{H}_m}$ only. Experiments favour the second model. Kaptein's calculations give the same results, for $J = 0$, $\omega = |\mathcal{H}_m|$ and $x_m \simeq \sqrt{|\mathcal{H}_m|}$. The diffusion model is now the only model used.

) **Extension to other Cases.** Our calculations are based on a simple encounter process. But such simple cases do not always occur, and some reactions may be combination of various encounter processes. KAPTEIN [81, 128a] has studied several cases.

i) Case of competitive reactions.
The general reaction sequence is:

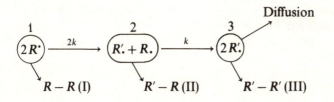

where the successive radical pairs with correlated spins are labelled 1, 2, 3. The
probability of a reaction for (I), (II), (III) can be calculated using the CKO theory
for S and T precursors, but the calculations are too lengthy to be given here.

ii) The case of stereospecific homolytic rearrangements.
The reaction scheme is:

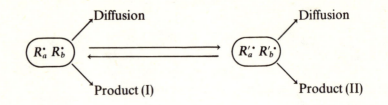

In this case the relation between the probability of formation of product I and
product II is given by KAPTEIN [81, 128b] for S and T precursors.

Recently, KESSENIKH et al. [206] have developed a similar theory, but they
use different rate constants for each different processes. They do not take into
account the exchange integral J and study the influence of a series of re-encounters.
From this last point they found the existence of a maximum of the CIDNP
effect. This theory was applied to the case of a two-level nuclear spin system and
to a methyl group.

2. Calculation of Enhancement Factors

a) c Products. The preceding theory gives us the probability of finding a c product
in the nuclear spin state χ_m. We may now deduce the populations of the different
NMR energy levels. Note that this population is generally quite different from
its normal value which proceeds from the Boltzmann distribution law. When the
reaction begins, the perturbation of the spectrum increases until it reaches a
stationary state, where the CIDNP effect is measured. The difference between a
perturbed intensity and its normal value when no reaction occurs gives the size
of the CIDNP effect. This is characterized by the enhancement factor V_{mn}, for the
transition between nuclear spin states m and n. This term was first introduced by
FISCHER [5]. Its value is determined by:

$$V_{mn} = \frac{I_{mn} - I^0_{mn}}{I^0_{mn}} \qquad (32)$$

where I^0_{mn} is the intensity of the transition $m \to n$ in the normal spectrum and
I_{mn} the intensity of the same line in the perturbed spectrum.

We have previously seen (2.1.5b) that the relative intensity of a transition is
given by (10):

$$I_{mn} = K' p_{mn}(N_m - N_n)$$

where p_{mn} comes from (9) with $\varphi_m = \chi_m$. As p_{mn} is the same in the two cases, we have:

$$V_{mn} = \frac{N_m - N_n}{N_m^0 - N_n^0} - 1 \tag{33}$$

where N_m^0 and N_n^0 are the numbers of spins in state m and n at thermal equilibrium, and N_m, N_n their value during the reaction. This result assumes that all relaxation times are equal; the influence of this parameter will be discussed later.

At thermal equilibrium, we can use Boltzmann's law and we have:

$$\frac{N_m^0}{N_n^0} = \exp\left(\frac{E_m - E_n}{kT}\right) \simeq 1 + \frac{E_m - E_n}{kT} + \cdots$$

where E_m and E_n are the energies of states χ_m and χ_n.

Then:

$$V_{mn} = \frac{(N_m - N_n)kT}{N_m^0(E_m - E_n)} - 1 \tag{34}$$

As usual, $I_{mn} \gg I_{mn}^0$ [23], so we can use an approximate value of V_{mn}:

$$V_{mn} \simeq \frac{I_{mn}}{I_{mn}^0} \quad \text{and} \quad V_{mn} = \frac{(N_m - N_n)kT}{N_m^0(E_m - E_n)}.$$

As nuclear energy levels are slightly different (2.1.5b) we may consider that in the unperturbed spectrum we have approximately the same number of molecules in each energy level. If we have $\mathcal{N}$ molecules and L nuclear energy levels, $N_m^0 = \mathcal{N}/L$. In the perturbed spectrum:

$$N_m = \frac{\mathcal{N} P_m}{\displaystyle\sum_{m=1}^{L} P_m}$$

where P_m is obtained from (29), (30) or (31).

The enhancement factor per molecule c formed is now:

$$V_{mn} = \frac{P_m - P_n}{\displaystyle\sum_{m=1}^{L} P_m} \cdot \frac{LkT}{(E_m - E_n)} - 1.$$

This enhancement factor has been described [16, 81, 128a] for cage products. With (31a) and (31b) we have for an S precursor,

$$V_{mn}^S = \frac{-(x_m - x_n)LkT}{\left(L\beta - \sum\limits_{m=1}^{L} x_m\right)(E_m - E_n)}. \tag{35}$$

If $x_m \ll \beta$

$$V_{mn}^S = \frac{-(x_m - x_n)kT}{\beta(E_m - E_n)} \tag{35 bis}$$

for a T precursor,

$$V_{mn}^T = \frac{(x_m - x_n)kT}{\left(\sum\limits_{m=1}^{L} x_m\right)(E_m - E_n)}. \tag{36}$$

KAPTEIN [81, 128a] has calculated the enhancement factor for the case of a thermal reaction and an S precursor:

$$V_{mn}' = \frac{I_{mn}}{I_{mn}^p(t) T_{1mn}^c k_{f_p}}$$

where I_{mn} has the same meaning as previously, $I_{mn}^p(t)$ is the intensity of the lines of the precursor at time t after the beginning of the reaction. This value is often difficult to obtain experimentally. T_{1mn}^c is the longitudinal nuclear relaxation time of transition mn of product c and k_{f_p} is the rate constant of formation of the pair from the precursor.

b) Diffusion Products: e Products. This problem was treated by KAPTEIN [81, 128a], CLOSS et al. [44, 45] and FISCHER [46].

We consider the process:

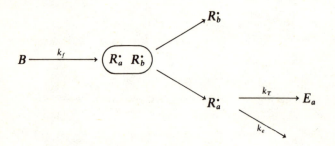

where E_a is an escape product.

The enhancement factor has been calculated and leads to the following expression:

$$\frac{I_{mn} - I_{mn}^0}{I_{mn}^0} = V_{mn}^e \, T_{1mn}^e \frac{k_T \, k_e}{(1/T_{1mn}^R + k_T + k_e)} \times f \tag{37}$$

with

$$V_{mn}^e = -\frac{(P_{ma} - P_{na})}{\sum_n (1 - P_n)} \cdot \frac{L_a k T}{(E_{ma} - E_{na})} \tag{37 bis}$$

where: T_{1mn}^e is the longitudinal relaxation time of the transition $m \to n$ of E_a, T_{1mn}^R that of radical $R_a^{\bullet}$, and k_e describes all other paths by which the radical $R_a^{\bullet}$ may disappear.

$f = \dfrac{r(t')}{\int\limits_0^{t'} r(t) dt}$ where $r(t)$ is the rate of the radical pair formation, L_a is the number

of the nuclear energy levels of product E_a, $P_{ma} = \sum\limits_{m_b} P_m$ where the sum is taken for

all nuclear spin states of radical b.

We then replace P_m by its value from (31a), (31b), or (31c) depending on the nature of the precursor.

The difference between c and e products appeared in Eqs. (34) and (37). The polarization has a different sign in the two cases.

3. Forms of the Spectrum. Kaptein's Rules

a) **First-Order Spectra.** For c products, the enhancement factor is proportional to $x_m - x_n$, as can be seen from (35 bis) and (36). As x_m is proportional to $\dfrac{\mathcal{H}_m^2}{\omega^{3/2}}$

with $\omega = \sqrt{J^2 + \mathcal{H}_m^2}$, if $J^2 \gg \mathcal{H}_m^2$, $\omega \sim J$ and $x_m \sim \mathcal{H}_m^2$. We have also $V_{mn} \simeq \dfrac{I_{mn}}{I_{mn}^0}$;

with $I_{mn} \simeq V_{mn} I_{mn}^0$ and as I_{mn}^0 is constant we have:

$$I_{mn} \sim \mathcal{H}_m^2 - \mathcal{H}_n^2 \,.$$

From (26) and for the transition between $m = [\dots M_i, M_j, M_k \dots]$ and $n = [\dots M_{i-1}, M_j, M_k \dots]$ of nucleus i of fragment a, one obtains:

$$I_{mn} \sim \tfrac{1}{2} A_i [\Delta g \beta_e H_0 + \sum_{j \neq i} A_j M_j - \sum_k A_k M_k + A_i (M_i - \tfrac{1}{2})] \,. \tag{38}$$

The j index refers to radical a and the k index to radical b of the radical pair $\boxed{a..b}$.

This expression was given by KAPTEIN [81, 128a] and LAWLER [113]. FISCHER [46] gave a similar expression without the last term.

The first term of (38) gives a net emission (E) or a net absorption (A) according to whether Δg is negative or positive. The sign of Δg is different for radical a and radical b because $\Delta g = g_a - g_b$ and we only consider fragment a. This effect is proportional to H_0 if $J^2 \gg \mathcal{H}_m^2$. If this condition is not satisfied, the problem is more complex.

The second term in (38) explains the multiplet effect which arises from coupling of nucleus i with the other nucleus j of the radical a, and it depends on the signs of A_i, A_j, M_j. This effect is also a function of J_{ij} which modifies the labelling of transitions [1c, 1e].

The third term of (38) contributes to the same effect between nucleus i of radical a and all the nuclei k of radical b. This interaction has an opposite phase compared to the second terms.

The last term is only of importance for second-order spectra. This question will be studied later.

KAPTEIN has also considered the case of e products [81, 128a].

b) **Kaptein's Rules** [79, 81, 128a]. The preceding calculations have shown that:

   i) the polarization from an S precursor is the opposite of those of T or F precursors;

   ii) the polarization of c products is the opposite of those of e products;

   iii) the net effect from $\Delta g A_i$ has a different sign for nuclei of fragment a and b;

   iv) the multiplet effect from $A_i^a A_j^a$ or $A_i^a A_k^b$ depends on the sign of J_{ij} and J_{ik};

   v) second-order multiplet effects may appear in the spectra of magnetically equivalent nuclei. This effect depends on the sign of J_{ij} but not on the sign of A_i since it is a function of A_i^2.

All these considerations, except for the last one, have been summarized by KAPTEIN in two equations, which give a qualitative prediction of the features for a high magnetic field:

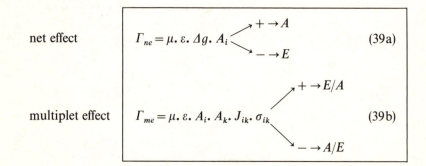

net effect $\qquad \Gamma_{ne} = \mu.\,\varepsilon.\,\Delta g.\,A_i \begin{cases} + \to A \\ - \to E \end{cases}$ (39a)

multiplet effect $\qquad \Gamma_{me} = \mu.\,\varepsilon.\,A_i.\,A_k.\,J_{ik}.\,\sigma_{ik} \begin{cases} + \to E/A \\ - \to A/E \end{cases}$ (39b)

In these relations we consider the nucleus i of fragment a of the radical pair. As defined in Chapter 1, A represents a net absorption, E a net emission, E/A and A/E represent a multiplet effect with conventions given in Chapter 1.

In these expressions:

$$\mu \begin{cases} + \text{ for } T \text{ and } F \text{ precursors} \\ - \text{ for a } S \text{ precursor} \end{cases}$$

$$\varepsilon \begin{cases} + \text{ for } c \text{ products} \\ - \text{ for } e \text{ products} \end{cases}$$

$$\sigma_{ik} \begin{cases} + \text{ if the indices } i \text{ and } k \text{ are of the same radical} \\ - \text{ if the indices } i \text{ and } k \text{ are of different radicals.} \end{cases}$$

Some applications will be given in the next chapter. When both net polarization and multiplet effect are present, the spectrum is the superposition of the two effects. But if $\Delta g \beta_e H_0$ is larger than A_i, Eq. (39b) can give erroneous results [79]. For strongly coupled spectra it is necessary to simulate the spectrum.

c) **Second-Order Spectra.** In this case the chemical shifts are of the order of J_{ij} [1a, 1c, 1e, 1f], and the spin functions χ_m are not eigenfunctions. The wave function of a nuclear state is then a linear combination ϕ of χ_m having the same total magnetic quantum number M as defined in 2.1.1.

$$\phi_l = \sum_m q_m^{(l)} \chi_m .$$

Then $c_S(t)$ and $c_{T_0}(t)$ given by (23) and (24) are not calculated with

$$\psi_m(t) = (c_{Sm}(t)\varphi_S + c_{T_0m}(t)\varphi_{T_0})\chi_m$$

of (22) but with

$$\psi_l'(t) = (c_{Sl}'(t)\varphi_S + c_{T_0l}(t)\varphi_{T_0})\phi_l \quad \text{or} \quad \psi_l'(t) = \sum_m q_m^{(l)} \psi_m(t) .$$

We then obtain:

$$c_{Sl}'(t) = \sum_m q_m^{(l)} c_{Sm}^{(l)}(t) \quad \text{and} \quad c_{T_0l}'(t) = \sum_m q_m^{(l)} c_{T_0m}^{(l)}(t) .$$

The only important quantity is:

$$|c'_{Sl}(t)|^2 = \sum_m q_m^{(l)2} |c_{Sm}^{(l)}(t)|^2$$

because χ_m are orthogonal functions.

The population of an energy level μ may be calculated with Eq. (27)

$$P_\mu = \lambda f(t)|c'_{Sl}(t)|^2$$

or

$$P_\mu = \sum_m q_m^{(l)2} P_m(t).$$

The enhancement factor $V_{\mu\nu}$ of a transition between energy levels μ and ν is then deduced from (34).

KAPTEIN and den HOLLANDER [81, 128a] have modified the LAME program. Using P_μ previously calculated, they have simulated CIDNP spectra.

d) Calculations of Intensities, Müller-Closs Rules. In general, the comparison between theory and experiment is made by simulation. MULLER [109] and CLOSS [83b] however have proposed a graphical method to determine intensities.

MULLER [109] has developed the A_3X case. (To understand this nomenclature see [1c, 1e, 1f])

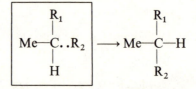

A refers to methylprotons and X to the neighboring proton. The normal spectrum of X is a quadruplet arising from a scalar coupling constant between X and protons of methyl A. The measured intensity I_{rel} of one line is represented as a function $G = \frac{1}{2}\beta_e H_0 \Delta g$ expressed in units of $A_A/4$, where A_A is the hyperfine coupling constant of the methyl protons.

Figure 2.5 shows such a diagram for: $|A_X| = |A_A|$ with $J = 0$

From this representation it is easy to find the general features of any other diagram describing A_nX systems by means of the following rules:

Rule 1: All curves have identical characteristics, are symmetrically displaced along the G axis with a constant interval of two diagram units, and the origin of the coordinates is a center of symmetry.

Rule 2: Their amplitudes are weighted by the binomial coefficients characterizing the relative line intensities of the corresponding multiplet at equilibrium and for first-order spectra.

Rule 3: To take into account the fact that A_X is generally different from A_A, MULLER [*109*] and CLOSS [*83b*] give the variation of $I_{ref}(G)$ with the parameter $p = A_X/A_A$ in the case $J = 0$. An example is given in Fig. 2.6.

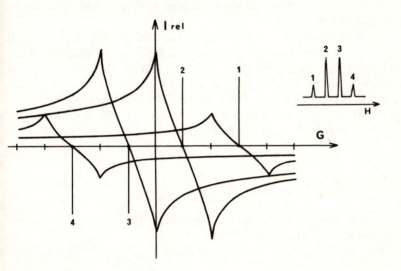

Fig. 2.5. Relative line intensities for a quadruplet of the X part of an A_3X spin system (From MULLER [*109*])

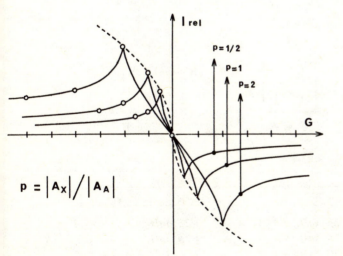

Fig. 2.6. Variation of I_{rel} with the ratio p for $J = 0$ (From MULLER [*109*])

Rule 4: The slopes of these curves at their zero point have been chosen as negative in Figs. 2.5 and 2.6. The sign of this slope is determined by the sign of the parameter $\varphi = \mu A_X \varepsilon$, where parameters have the same meaning as in (39a) and (39b).

Rule 5: Each curve must be associated with a particular line in the multiplet. If the sign of parameter l defined by: $l = A_A \cdot J_{AX} \cdot \sigma_{ik}$ is positive, lines are counted from low to high field as in Fig. 2.5; if it is negative, lines are counted in the opposite direction.

J_{AX} in l is the scalar coupling constant between protons A and X. Examples have been given by CLOSS [*83b*].

If J is not zero, there are no significant modifications for $|J| \langle 0.1 | A_A |$. If $|J| \rangle |A_A|$, curves are straight lines over an extensive region of the diagram. If J is of the same order, the maximum of the curve is not so sharp. CLOSS [*83b*] (p. 47) publishes diagrams for $J = |A_A|/2$ and $J = |A_A|$.

These rules may be extended to $A_n X_m$ spectra. The diagram of the X part is obtained from an $A_n X$ case and each curve is the sum of N components, each one being built up as before except for Rule 3.

4. Example

We consider two protons with different chemical shifts. This is an AX case where J_{AX} is the spin-spin coupling constant.

The labelling of the line is different according to the sign of J_{AX}, as shown in Fig. 2.7a.

The following table contains the value of $\mathcal{H}_m$ calculated with (26) in two cases for each function χ_m. In case A, there is one proton on each radical of the radical pair. In case B, the two protons are on the same radical.

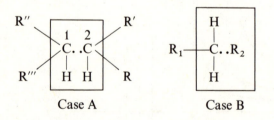

Case A   Case B

χ_m	Case A	Case B
$\beta\beta$	$1/2\beta_e \Delta g H_0 - A_A/4 + A_X/4$	$1/2\beta_e \Delta g H_0 - A_A/4 - A_X/4$
$\alpha\beta$	$1/2\beta_e \Delta g H_0 + A_A/4 + A_X/4$	$1/2\beta_e \Delta g H_0 + A_A/4 - A_X/4$
$\beta\alpha$	$1/2\beta_e \Delta g H_0 - A_A/4 - A_X/4$	$1/2\beta_e \Delta g H_0 - A_A/4 + A_X/4$
$\alpha\alpha$	$1/2\beta_e \Delta g H_0 + A_A/4 - A_X/4$	$1/2\beta_e \Delta g H_0 + A_A/4 + A_X/4$

Example   55

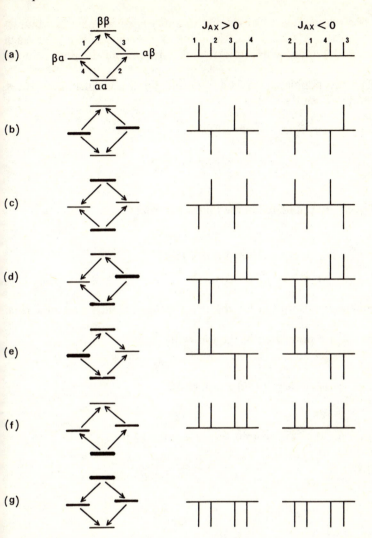

Fig. 2.7. CIDNP effect on an AX spin system (see text) (From CLOSS [83b])

We assume that $A_A \sim A_X > 0$ and we consider only the case of c products. For e products the result may be inverted as predicted by Kaptein's rules. We call $^{\chi_m}\mathscr{H}$ the coefficient of the upper table for the state χ_m.

i) $\Delta g = g_1 - g_2 = 0$

In case A we have:

$$0 \simeq |^{\alpha\alpha}\mathscr{H}| = |^{\beta\beta}\mathscr{H}| < |^{\alpha\beta}\mathscr{H}| = |^{\beta\alpha}\mathscr{H}|.$$

T precursor: The overpopulated levels are $\alpha\beta$ and $\beta\alpha$, so we have a multiplet effect A/E for $J>0$ and E/A for $J<0$. The result is presented in Fig. 2.7.b, each level is represented by a line whose thickness is correlated to the population of this state.

S precursor: The overpopulated levels are $\alpha\alpha$ and $\beta\beta$, a multiplet effect arises, E/A for $J>0$ and A/E for $J<0$ (Fig. 2.7c).

In case B we have:

$$|{}^{\beta\beta}\mathcal{H}| < |{}^{\alpha\beta}\mathcal{H}| \simeq |{}^{\beta\alpha}\mathcal{H}| < |{}^{\alpha\alpha}\mathcal{H}|\,.$$

T precursor: Levels are populated in the following order:

$$(\beta\beta) < (\alpha\beta) \simeq (\beta\alpha) < (\alpha\alpha)\,.$$

We have an enhanced absorption for the two protons for all values of J (Fig. 2.7f).

S precursor: Levels are populated in the following order:

$$(\alpha\alpha) < (\beta\alpha) \simeq (\alpha\beta) < (\beta\beta)\,.$$

We have emission for the two protons for all J (Fig. 2.7g).

ii) $\frac{1}{2}\beta_e \Delta g H_0 = \frac{1}{2}\beta_e(g_1 - g_2)H_0 \gg A_A \simeq A_X > 0\,.$

In case A we have:

$$|{}^{\beta\alpha}\mathcal{H}| < |{}^{\alpha\alpha}\mathcal{H}| \simeq |{}^{\beta\beta}\mathcal{H}| < |{}^{\alpha\beta}\mathcal{H}|\,.$$

T precursor: Levels have different populations:

$$(\beta\alpha) < \beta\beta \simeq (\alpha\alpha) < (\alpha\beta)\,.$$

We have an enhanced absorption for the proton on carbon C_1 and an emission for the proton on C_2 for all values of J (Fig. 2.7d).

S precursor: Levels are populated in the following order:

$$(\alpha\beta) < (\alpha\alpha) \simeq (\beta\beta) < (\beta\alpha)\,.$$

We have an emission for the proton on C_1 and an enhanced absorption for the proton on C_2 for all values of J (Fig. 2.7e).
In case B we have:

$$|^{\beta\beta}\mathscr{H}| < |^{\alpha\beta}\mathscr{H}| \simeq |^{\beta\alpha}\mathscr{H}| < |^{\alpha\alpha}\mathscr{H}|.$$

T precursor: We then have, for the different populations:

$$(\beta\beta) < (\alpha\beta) \simeq (\beta\alpha) < (\alpha\alpha).$$

We have an enhanced absorption for all protons (Fig. 2.7f).
S precursor: Levels are populated in the following order:

$$(\alpha\alpha) < (\beta\alpha) \simeq (\alpha\beta) < (\beta\beta).$$

We have an emission for all protons (Fig. 2.7g).

Other examples have been given in the literature, for instance those of WARD and CHIEN [56] for the system CH_2—CH=CH_2.

Practical examples of AX cases have been given by CLOSS [83b] (p. 89) and by SHINDO [96].

IV. Kinetic Formulation of the Radical-Pair Mechanism

FISCHER [46, 76, 102a] and MARUYAMA et al. [95] give an interpretation of CIDNP using the radical-pair model, but with a kinetic treatment which shows some particular aspects of the phenomenon which are not evident in the CKO theory.

1. Principle

Nuclear polarizations are attributed to adiabatic transitions between the non-adiabatic singlet and triplet state of the radical pair arising from geometrical fluctuations.

The mixing of S and T states occurs for inter-radical distances whose values are situated in zone II of Fig. 2.4. This mixing arises from the hyperfine coupling constant A [Eq. (5)] and from the difference between the electron Zeeman levels which arises from the interaction ($\omega_s = \Delta g \beta_e H_0$) of the radical electrons in the magnetic field H_0. This mixing between S and T_0 explains the difference in population of the nuclear energy level.

We shall consider a radical pair consisting of two radicals R_1 and R_2 with Λ_1 nucleus of spin I_{λ_1} and Λ_2 nucleus of spin I_{λ_2}, respectively. The number of the nuclear energy levels of the pair is:

$$n = \prod_{\lambda_1=1}^{\Lambda_1} (2I_{\lambda_1} + 1) \prod_{\lambda_2=1}^{\Lambda_2} (2I_{\lambda_2} + 1).$$

A total-spin function is obtained as before from the product of an electron-spin function and a nuclear-spin function. Then we have $n(S$ states$)$ and $3n(T$ states$)$. If we consider only T_0 states we have n (T states). The singlet will be labelled i and the triplet j_σ with $\sigma = 0 \pm 1$. The diffusion model of NOYES, gives n_c transitions in zones I and II of Fig. 2.4 during the evolution of the radical pair. The kinetic model of FISCHER uses the following rate constants:

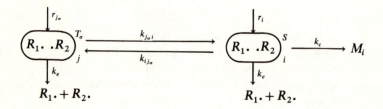

The probability of finding a pair in a selected initial state after n_c librations is given by:

$$P_{i\,j_\sigma}^{\text{total}} = 1 - 2n_c P_{ij_\sigma}(1 - P_{ij_\sigma}) \simeq 1 - 2n_c P_{ij_\sigma}.$$

Transitions between states i and j_σ occur with the rate constants $k_{ij_\sigma} = -P_{ij_\sigma}^{\text{total}}$ and and $k_{j_\sigma i}$ respectively during the lifetime of the pair. Escape products are formed with a rate constant $k_e = k_{ei} = k_{ej_\sigma}$ and reaction occurs with a rate constant k_c. If we suppose that products are formed only from S states, then $k_{ci} = k_c$ and $k_{cj_\sigma} = 0$. The rate equations for the population of singlet states and triplet states are:

$$dN_i/dt = r_i - \left(k_c + k_e + \sum_\sigma \sum_j k_{ij_\sigma}\right) N_i + \sum_\sigma \sum_j k_{j_\sigma i} N_{j_\sigma} \tag{40a}$$

$$dN_{j_\sigma}/dt = r_{j_\sigma} - \left(k_e + \sum_i k_{j_\sigma i}\right) N_{j_\sigma} + \sum_i k_{ij_\sigma} N_i \tag{40b}$$

where N_i or N_{j_σ} is the population of the corresponding state i or j_σ and r_i, r_{j_σ} are the pair production rates in states i or j_σ.

As reaction products are formed with S states, the number of molecules produced in nuclear-spin state i are:

$$dM_i/dt = k_c N_i = \dot{M}_i. \tag{41}$$

Free radicals in nuclear-spin state i which escape the radical pair are produced at a rate:

$$\dot{R}_{ei} = dR_{ei}/dt = k_e \left(N_i + \sum_\sigma N_{j_\sigma} \right). \tag{42}$$

The enhancement factor between states i and i' is:

$$V_{ii'} = \frac{\dot{M}_i - \dot{M}_{i'}}{\dot{M}_i + \dot{M}_{i'}} \cdot \frac{1}{\langle I_{ii'} \rangle_0}. \tag{43}$$

The solution of the simultaneous equations (40a), (40b), (41) introduced into (43) gives the enhancement factor.

2. Application

FISCHER [46] considers the case where the population state of the pairs is time-independent and has only one nuclear spin state, $i = 1$ and $j_\sigma = 1$. The sum over j or i then disappears. Two subcases are considered.
a) Nuclear spin configuration does not change as a transition occurs, i.e. $j = 1$ and $k_{ij_\sigma} = k_{j_\sigma i}$. This corresponds to the case where anisotropic interactions are neglected.
  i) for a S precursor $r_i = r/n$, $r_{j_\sigma} = 0$,
  ii) for a T precursor $r_i = 0$, $r_{j_\sigma} = r/3n$,
  iii) for a F precursor $r_i = r_{j_\sigma} = r/4n$.
The expression for N_i is given [46].

From these calculations, the following rules are deduced, although these results may also be obtained from previous theories.
1. The sign of the enhancement factor for c products is different from that of the e product for the same NMR transition and the same precursors.
2. The sign of the enhancement factor for a S precursor is different from that of T or F precursors for the same NMR transition and the same products.
3. For equal product formation rate, the enhancement factor is greater for T precursors than for F precursors.

b) Nuclear spin configuration is not conserved ($j_\sigma \neq i$). The simple case is then considered: one singlet state i will mix with only one triplet state j_σ; the corresponding j_σ state will not mix with a singlet state. This is the case for $S\text{-}T_-$ transitions.

From (42) the same conclusion as in a) may be obtained except for (1) where the signs are identical. Previously in a) we have had no CIDNP effect if no pairs were formed: $k_c = 0$ or, if we do not have pair escape, $k_e = 0$. Here no CIDNP effect is observed if $k_e = 0$ and $k_{ij_\sigma} = 0$, but there is a maximum polarization for $k_c = 0$. No experimental example is known at present for this case.

3. k_{ij_σ} Calculation

This is a very tedious calculation and FISCHER [46] gives an approximate treatment. Exact calculation is not possible. The transition probability between continuous states obtained by COULSON and ZALEWSKI is used to estimate P_{ij_σ}.

$$2P_{ij_\sigma} = 1 - \exp\left\{\frac{2/v^2 |\int \Omega_{ij_\sigma} dR|^2}{1 + 1/v^2 |\int \Omega_{ij_\sigma} dR|^2}\right\}$$

where Ω_{ij_σ} is the matrix element of the electronic part of the Hamiltonian between state i and j_σ, v is the classical velocity dR/dt where R is the distance between the two radicals of the radical pair.

As transitions occur in a small area ΔR, we have approximately:

$$|\int \Omega_{ij_\sigma} dR|^2 \simeq \Omega_{ij_\sigma}^2 \Delta R^2$$

then,

$$k_{ij_\sigma} = \frac{dP_{ij_\sigma}^{\text{total}}}{dt} \simeq (k_c + k_e) 2n_c P_{ij_\sigma} = 2n_c(k_e + k_c)\frac{\Delta R^2}{v^2}|\Omega_{ij_\sigma}|^2$$

J of Eq. (15) has the form:

$$J = J_0 e^{-R/R_0}.$$

Then

$$|\Delta R| \simeq R_0 \frac{|\Omega_{ij_\sigma}|}{J_M} \qquad \text{with} \qquad |J| \simeq |\Omega_{ij_\sigma}|$$

and J_M is the value of J in the mixing region.

Starting from $\mathcal{H}$ given by (14) and after a discussion about the magnitude of each terms, FISCHER found for a transition $T_0 - S$ and two identical nuclear spin states:

$$|\Omega_{ij_0}|^2 = |\Omega_{j_0 i}|^2 = 1/4 \left| \frac{\beta_e \Delta g H_0}{\hbar} + \sum_{\lambda_1} A_{\lambda_1} m_{I_{\lambda_1}} - \sum_{\lambda_2} A_{\lambda_2} m_{I_{\lambda_2}} \right|^2 .$$

This expression is very similar to $\mathcal{H}_m^2$ given in (26).

These calculations give the same quantitative results as the CKO theory but the equations are not easy to handle.

Other methods especially those using the density matrix formulation [117, 171] will be studied in 2.5 because they are more general and may be applied to the case of low magnetic fields.

V. The Influence of the Magnetic Field H_0 on the CIDNP Spectra

1. Introduction

The influence of the strength of the magnetic field was first mentioned by WARD et al. [26]. The reaction began outside the magnet and then CIDNP was observed in an NMR spectrometer in a high field. This influence was also mentioned by LEHNIG and FISCHER [22, 149], RYKOF [40], GARST [55], LAWLER [102q], and SAGDEEV et al. [208, 209]. Figure 2.8 gives an example of this effect.

If the magnetic field is low, the main difference from the situation previously studied is that mixing with T_+ and T_- now occurs in zone II of Fig. 2.4. The energy difference between $T_\pm$ and T_0 is a function of H_0 and is very small. We may now introduce T_+ and T_- in the former calculations, especially in $\psi_m(t)$ of Eq. (22).

This new theory has been considered in several different ways. ADRIAN [104] and CHARLTON et al. [89] introduced $T_\pm$ and developed calculations which are

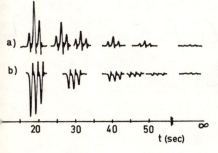

a)

b)

20   30   40   50   ∞

t (sec)

Fig. 2.8. NMR spectra of $C_6H_5CHClCH_2CH_3$ produced from the reaction of α, α-dichlorotoluene with ethyl lithium. The time axis indicates the number of seconds after the reagents were mixed. A – Reagents were mixed in the spectrometer probe, B – Reagents were mixed outside the spectrometer field and placed in the probe 12 sec after mixing (From WARD et al. [26])

complicated. MORRIS et al. [117] introduced the density matrix formulation and gave a general interpretation of the CIDNP spectra. On the other hand KAPTEIN [81, 129] has discussed the case where the reaction occurs outside the magnet or in a low magnetic field and CIDNP is observed in an NMR spectrometer. GLARUM [176a] has also considered this case. We shall now relate these different treatments.

2. Extension of the CKO Theory

ADRIAN [104] has considered the general problem, SHINDO et al. [96] and CHARLTON et al. [89] have limited their investigations to the case of two protons.

The wave function (22) is now written:

$$\psi_{\pm m}(t) = (c_{T \pm m}(t)\varphi_{T \pm} + c_{S \pm m}(t)\varphi_S)\chi_m . \tag{44}$$

Calculations are then made with T_+ or T_- only. Equations (44) are solutions of the Schrödinger equation:

$$dc_{T\pm}/dt = -i[\pm 1/2g\beta_e H_0 + J(t)]c_{T\pm} - iac_{S\pm} \tag{44a}$$

$$dc_{S\pm}/dt = -iac_{T\pm} + i[\pm g\beta_e H_0 + J(t)]c_{S\pm} . \tag{44b}$$

In these equations, the reference energy is equal to $\pm g\beta_e H_0$. Energy is expressed in units of $\hbar$. $2J(t)$ is the energy difference between singlet and triplet states. $\pm g\beta_e H_0$ is the triplet state electron Zeeman energy and a the matrix element of the magnetic interaction connecting the singlet and the triplet states. These matrix elements may be the hyperfine interaction $(A_1 S_1 . I_1 + A_2 S_2 . I_2)$ (for one nucleus of spin $\frac{1}{2}$: $a^2 = |\langle T_- \alpha | A_1 S_1 . I_1 | S\beta \rangle|^2 = |\langle T_+ \beta | A_1 S_1 . I_1 | S\alpha \rangle|^2 = A^2/8$) or spin rotation interaction, as seen in 2.1.4.b. The solution of (44a) and (44b) is given by ADRIAN [104]. One can see that $T_+ \to S$ decreases the number of electron spins parallel to H_0 and, if a hyperfine interaction mixes the singlet and triplet, the number of nuclear spins parallel to H_0 increases. The opposite is the case for the $T_- \to S$ transition.

Since reaction products are formed only from a radical pair in an S state, if the pair is initially in a T state, the probability of reaction is:

$$P(t) = |c_{S_+}(t)|^2 + |c_{S_-}(t)|^2 .$$

From detailed calculations, two features appear:
  i) There is no triplet singlet mixing if

$$J^2 \gg a^2 + (\tfrac{1}{2}g\beta_e H_0)^2 .$$

  ii) When triplet-singlet mixing occurs, a time of the order of $[a^2 + (\tfrac{1}{2}g\beta_e H_0)^2]^{1/2}$ is required for the development of polarization. For classical values of a and H_0, this time is longer than the time between individual diffusive displacements of the radicals. As J decreases rapidly with the distance r between radicals, it is impossible to have a large polarization.

  To overcome this difficulty ADRIAN [104] suggests the following sequence of encounters in the case when the radicals initially fail to recombine. If t_1 and t_2 are the time of first and second encounters it follows:

$$0 \leq t \leq t_1 \longrightarrow J = 0$$
$$t_1 \leq t \leq t_2 \longrightarrow J^2 \gg a^2 + (\tfrac{1}{2}g\beta_e H_0)^2$$
$$t_2 \leq t \leq t_3 \longrightarrow J = 0 .$$

If we start from a triplet state: $|c_{T_{\pm}}(0)|^2 = \tfrac{1}{3}$.
The probability of reaction for the third encounter is:

$$P_T(t_3) = -\left(g\beta_e H_0 a^2 \frac{J}{3} |J|\omega^3\right) \times [\sin 2\omega t_1 \sin 2|J|(t_2 - t_1) \sin^2 \omega(t_3 - t_2)$$
$$+ \sin^2 \omega t_1 \sin 2|J|(t_2 - t_1) \sin 2\omega(t_3 - t_2)] \tag{45}$$

with

$$\omega = \sqrt{a^2 + (\tfrac{1}{2}g\beta_e H_0)^2} .$$

If we start from a singlet state the sign is reversed.
  Equation (45) and the diffusion model of NOYES [4] (see 2.1.6.b) in conjunction with (12) give the mean value of P_T:

$$P_T = -0.014 \left(\frac{g\beta_e H_0 a^2}{\omega^3}\right) \sqrt{\omega \tau_D}$$

where τ_D is the time interval between diffusive displacements of the radicals.

We see that there is only an emissive proton NMR spectrum: $\langle P_T \rangle < 0$, for a radical pairs initially in a triplet state.

Detailed calculations for the case of two protons have been made by CHARLTON and BARGON [89]. For the electronic states we have $T_\pm$, T_0 and S, and for nuclear states of the two protons we can also use $t_\pm$, t_0 and s states constructed as for two electrons.

As before, the total spin function of the radical pair is the product of one electron-spin function and one nuclear-spin function. We have 16 different spin states. From the theory of CHARLTON [89] it follows that only twelve off-diagonal elements are not equal to zero for electronic transitions between T and S states. Since the final state for a reaction must be a singlet state, the only off-diagonal elements of interest are:

$$T_+ t_0 | S t_+, \quad T_+ t_- | S t_0, \quad T_+ t_- | S s, \quad T_+ s | S t_+,$$

$$T_- t_+ | S t_0, \quad T_- t_+ | S s, \quad T_- t_0 | S t_-, \quad T_- s | S t_-,$$

$$T_0 t_+ | S t_+, \quad T_0 t_0 | S t_0, \quad T_0 t_- | S t_-, \quad T_0 s | S s.$$

The last four terms are those met in 2.3.4 and the others explain $T_\pm \to S$ transitions.

As we only have transitions between states having an off-diagonal element, we can see that the total magnetic quantum number $M = M_e + M_N$ remains constant during the transition.
For instance:

$$T_+ t_0 \longrightarrow S t_+$$
$$\downarrow \qquad\qquad\qquad \downarrow$$
$$M = 1 + 0 = 1 \quad M = 0 + 1 = 1$$

This is a general rule.

Consequently, for a high magnetic field where only $T_0 \to S$ transitions occur for electrons, nuclear-spin states do not change in the CKO theory. Now for $T_\pm \to S$ transitions nuclear-spin states are mixed. For instance, a t_+ nuclear spin state is populated via $T_+ t_0 \to S t_+$ and $T_+ s \to S t_+$ transitions. Then the transition probabilities can be calculated and this theory permits the interpretation of the CIDNP effect in a low field [89, 55].

3. The Density-Matrix Formulation (GCKO Theory)

This is the most general way to interpret the CIDNP effect. Only MORRIS et al. have outlined this method [117]. A good introduction to density matrix formulations for NMR is given by SCHLICHTER [1d] (p. 127). In this theory, we do not consider one radical pair as before, but an ensemble of N radical pairs. If γ_k is the probability that a radical pair chosen at random gives rise to a product in the nuclear spin state χ_k, then the number of molecules in state χ_k is: $n_k = \gamma_k N$. The elements $\gamma_{k'k}$ of the density matrix are expressed on the basis $\varphi_{Xi} = \phi_X \chi_i$ where ϕ_X is the spin function of T_+, T_0, T_-, S states, and χ_i is the ith nuclear-spin state. On this basis γ_k is a diagonal element. The density matrix of the ensemble of N radical pairs is $P(t)$. The formalism of this theory is then applied to compute $P(t)$ from an initial state $P(0)$. Detailed calculations are given in Reference 117. The theory gives n_k, from which it is possible to calculate the enhancement factor V_{mn} defined in (32):

$$V_{mn} = \frac{n_m - n_n}{n_m^0 - n_n^0} - 1 = \frac{4.89 \ 10^6 (T/H_0) 2^P (n_m - n_n)}{\sum_k n_k} - 1 \qquad (46)$$

where n_n is the population of the lower energy level, P is the number of protons and T the temperature of the sample in the NMR probe.

Application is made first for S and T precursors which have a well-defined initial $P(0)$. The initial populations of states are assumed to be equal. For F precursors, Kaptein's model [81, 128a] and 2.3 is used. In this case, radical pairs in an S state are assumed to react during their initial collision.

As the calculations are tedious, computer programs have been written. They have been called GCKO and NIMCO. Two cases are analysed.

a) Radical pairs with one proton.

Two subcases are calculated, one with $A = -22$ Gauss and one with $A = +27$ Gauss. The enhancement factor V_{mn} is given as a function of the magnetic field strength. Different values of J are considered. It is proved that V_{mn} shows considerable dependence on J, A and τ where τ is the mean radical-pair lifetime.

Another interesting feature of this calculation is the determination of the different probabilities for different possible pathways. For one proton there are four possible pathways: $T_0\alpha \to S\alpha$, $T_0\beta \to S\beta$, $T_+\beta \to S\alpha$ and $T_-\alpha \to S\beta$. Graphical results, as reported in Fig. 2.9 for example, are given for this case with fixed parameters [117].

These figures show the behavior of the population at each nuclear energy level. It is also clear that the premise $J = 0$, as previously used by ADRIAN [104], during diffusion processes is not always correct, but his result is the same.

b) Radical pairs with two protons.

The program is used to compute the field dependence of the CIDNP spectrum of the reaction $\boxed{Cl_2CH..CHClCOOH} \to Cl_2CH-CHClCOOH$ as previously studied by FISCHER and LEHNIG [75]. Figure 2.10 shows the agreement between the result obtained with this program and the experiment. Ordinates are proportional to V_{mn} of (46) where $\Delta n = n_m - n_n$ and $P = 1$.

This method seems to be the best for calculating the CIDNP effect. It enables the calculation of spectra for all H_0 values, even for the high-field case which was first explained using the CKO method.

Very recently EVANS et al. [171] have developed a general theory of CIDNP and CIDEP for high field using the density-matrix formalism. It is similar to that proposed by PEDERSEN [170l, 170s] for CIDEP and presented in Chapter 4. The major feature of this theory is that J is not supposed to be constant. The motion of the different spins in the radical pair is calculated using the stochastic Liouville method which leads to the calculation of the spin-density matrix. To solve their equation, they distinguish two types of interactions: short range for spin-spin interactions between the two radicals and long range for the motion of the two radicals. They also include NMR and EPR relaxation processes. The exchange

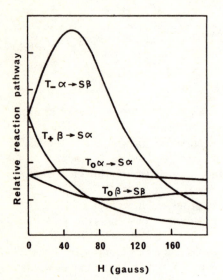

Fig. 2.9. Relative reaction pathway versus H_0 for a one-proton radical pair (c product) (From MORRIS et al. [117])

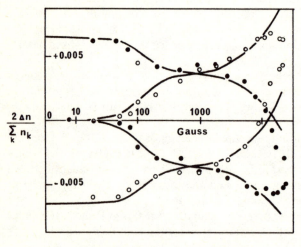

Magnetic Field of the Reaction

Fig. 2.10. Comparison between experimental data (circles) and values predicted by GCKO theory (see text) (From MORRIS et al. [117])

integral J is supposed to interact in a small domain (region II, Fig. 2.4) corresponding to a distance a between the two radicals. J is equal to: $J(r) = J \delta(r-a)/4\pi a^2$ where δ is the Dirac function. They explain the CIDNP effect with the only constants: the exchange integral J, the rate constant of recombination K, the initial distance of the two radicals r_0 and the distance a. All others parameters may be determined by other technics.

4. Kaptein's Theory for Low Field

KAPTEIN [81] and KAPTEIN and DEN HOLLANDER [129] calculate the enhancement factor in the case where the reaction occurs in a low magnetic field H_r and where the observation is performed in a high field. The computation is divided into three steps.

i) The Hamiltonian matrix for the radical pair in H_r is computed and then diagonalized.

ii) When the eigenvalues and eigenvectors have been calculated in the first step, the populations of the nuclear levels of the products are evaluated by means of the CKO model.

iii) From these populations in H_r, the population in a high field H_0 is then determined.

During the passage from H_r to H_0, the population may be transferred from the low-field energy level diagram to one of high field. Calculation in low-fields uses $T_\pm$ states mixing which gives more complicated equations. Results are given by KAPTEIN for c and e products with S, T and F precursors in the magnetic field H_r.

There is one difficulty in the passage from H_r to H_0. Usually spectra observed with low field are second-order spectra and those with high field are first-order. During this passage, energy-level diagrams are severely disturbed. For simple spectra such as $A_n B$, the correlation between the two energy-level diagrams may be easily found. This is not the case for more complex spectra. Such a correlation

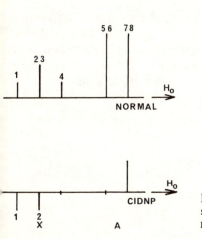

Fig. 2.11. The "n-1 multiplets" on an $A_2 X$ spin system (From KAPTEIN [81], and KAPTEIN and DEN HOLLANDER [129])

between the two energy-level diagrams is lost during the diagonalization procedures. To establish this correlation, intermediate spectra for field strength between H_r and H_0 are calculated, but this procedure is time-consuming. KAPTEIN [81, 129] mentioned 300 steps for propene from $H_r = 0.5$ Gauss to 14,000 Gauss.

Nevertheless, from this theory the variation of the enhancement factor V_{mn} with magnetic field strength is established. The results are the same as seen in the preceding section and in reference [117].

The zero field case is discussed in some detail. For one proton there is no CIDNP effect. Polarization can only arise when at least two nuclei with different chemical shifts are coupled. Figure 2.11 from [81, 129] shows the zero field effect on an A_2X case where $J_{AX} > 0$, $A_A > 0$ and $A_X < 0$.

Several lines have disappeared (one in each multiplet). This illustrates the so-called "$n-1$ multiplets" effect, which may be predicted by the simple Rule (39b) of this chapter.

5. Glarum's Theory

GLARUM [176a] considers the mixing between S and one triplet state. He studies first the nuclear polarization found in radicals formed from the dissociation of a molecule in a singlet electronic state (S precursor). The main difference with previous theories is the decomposition of the hamiltonian $\mathscr{H}$.

$$\mathscr{H} = \mathscr{H}_0 + \mathscr{H}'_{iso} + \mathscr{H}'_{aniso}$$

where

$$\hbar^{-1}\mathscr{H}_0 = \tfrac{1}{2}(g_1 + g_2)\beta H_0(Sz_1 + Sz_2) + J\mathbf{S}_1.\mathbf{S}_2 + \mathbf{S}_1.\mathbf{D}.\mathbf{S}_2$$

$$\hbar^{-1}\mathscr{H}'_{iso} = \tfrac{1}{2}(g_1 - g_2)\beta H_0(Sz_1 - Sz_2) + A_1\mathbf{S}_1.\mathbf{I}_i + \sum_j A_j\mathbf{S}_1.\mathbf{I}_j - \sum_k A_k\mathbf{S}_2.\mathbf{I}_k$$

$$\hbar^{-1}\mathscr{H}'_{aniso} = \mathbf{S}_1.(A_i - a_i\varepsilon)\,\mathbf{I}_i + \cdots.$$

This last term introduce a new term in the calculation of the CIDNP effect. But it is not very important.

The theory is then extended to geminate combination and/or disproportionation products. $\mathscr{H}'_{aniso}$ introduces also an other term not found in previous theories.

The case of radical transfer reactions is outlined using a three electrons-radical pair as GERHART et al. [136]:

$$AX + B. \rightleftarrows \boxed{\begin{array}{c} .X. \\ A.\ B \end{array}} \rightarrow A. + BX$$

but calculations are too complex to be made.

This theory is used to explain the influence of the magnetic field on the CIDNP effect. The radical pair is studied out of the magnet and then the intensity of lines are calculated in the high field case when the sample is in the spectrometer. A detailed discussion of the different forms of $f(t)$ is given, where it is shown that the two proposed forms: $f(t) \sim t^{-3/2}$ or $f(t) \sim e^{-t/\tau}$ are not general. Further experiments are required to improve this function.

VI. Relaxation and other Effects

1. Relaxation

As mentioned in 2.3.2.a, all previous calculations assumed that the longitudinal relaxation times T_1 of all transitions are the same. This is not always the case. Several authors [9, 16, 17, 23, 44, 45, 76, 81, 83b, 114, 128a, 149, 171] have mentioned the possible effect of relaxation times. One can distinguish two T_1 values: T_1 occurring in NMR spectra of products, which are of the order of 1 to 30 sec, and T_{1R} occurring in free radicals, which are in the range 10^{-3} to 10^{-5} sec. This shortening is due to the large dipole-dipole interaction between the free electron and the nucleus.

An example of the first kind is given by MULLER and CLOSS [114] in the X part of an $A_3 X$ system. Figure 2.12 shows the effect on these spectra: a) all T_1 are equal, b) with relaxation due to intermolecular interactions (T_1 inter), c) with relaxation arising from intramolecular interactions (T_1 intra). As can be seen in this example, the effect may be very large. In Eq. (10) we have assumed that the variation of the population N_m and N_n arises only from products having a nuclear state with an energy E_m or E_n and from the transition between E_m and E_n. This is true if all relaxation times T_1 are equal for all the transitions. More generally

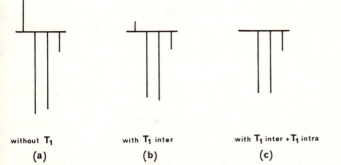

without T_1 with T_1 inter   with T_1 inter + T_1 intra

   (a)   (b)   (c)

Fig. 2.12. The effect of nuclear relaxation time on an $A_3 X$ spin system (X part) (From MULLER and CLOSS [114])

we have:

$$\frac{dI_{mn}}{dt} \simeq \frac{d(N_m - N_n)}{dt} = k_r[M](w_m - w_n) - 2w_{mn}(N_m - N_n) + \sum_{i \neq m} w_{im}(N_i - N_m)$$
$$- \sum_{i \neq n} w_{in}(N_i - N_n)$$

where w_n or w_m are the nuclear spin-state dependent probabilities of product formation from the pair, and w_{ij} the transition probability between level i and level j. If T_1 is the same for all transitions, the last two terms vanish, but if not, differences appear which modify the intensities of the spectrum.

In the stationary state $dI_{mn}/dt = 0$ and with the various theories which relate w_{ij} to the different relaxation times T_1 inter and T_1 intra, MULLER et al. [114] obtained the results shown in Fig. 2.12. These results allow us to interpret the photoinduced α cleavage of phenyl α phenylethyl ketone I from the radical pair II.

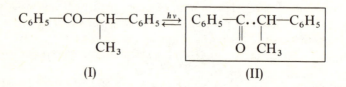

$$(I) \qquad\qquad\qquad\qquad\qquad (II)$$

A very good test of this theory is the use of the "spin-tickling" experiment [1e, 1f]. A transition is slightly irradiated with a second radiofrequency field and the populations of the energy levels connected by the irradiated transition become equal. It is easy to calculate the new spectrum which will be observed if such a perturbation is applied. MULLER [114] gives an example of such an experiment which shows good correlation with the theory and points out the important effect of relaxation phenomena.

Relaxation of the second kind (nuclear relaxation in free radicals) has been observed by CLOSS and TRIFUNAC [44], WARD, LAWLER and COOPER [25] and LEPLEY and LANDAU [29]. In this case T_{1R} is of the order of the lifetime τ_c of the radical pair. Consequently, if $T_1 < \tau_c$ the polarization may be lost. Such a situation has been discovered by CLOSS [44, 83b] where a product is formed with no polarization.

2. Sample Form

In all cases, reactions are produced in the cylindrical tubes used in NMR spectrometers. But DEUTCH [123] has studied the theoretical case of thin films in the high field H_0. In Eq. (27) we mentioned $f(t)$ which describes the motions of radicals in the radical pair. DEUTCH proposes to modify this function. As temperature or density will not change the process sufficiently, the only means of changing $f(t)$ is by the use of thin films. We then have CIDNP in a two-dimensional space.

This is difficult to achieve experimentally, but if the thickness of the sample is gradually decreased, the special properties of the thin film may progressively appear.

The calculations of $f(t)$ are performed with a continuous model by means of diffusion equations. He obtains an expression of the form:

$$f_3(t) = A(t)^{-3/2} e^{-\alpha/t}$$

for a three-dimensional space. When $t \to \infty$, $f_3(t) \sim t^{-3/2}$, which is equivalent to Eq. (13) used by KAPTEIN and CLOSS. From the expression $f_3(t)$, the probability of reaction for c products with T and S precursors is calculated and agrees with results in 2.3.

The calculations are then made for the two-dimensional space. This is very complicated, because some integrals cannot be calculated analytically. Only the asymptotic behavior of $f_2(t)$ is given:

$$f_2(t) = \left(\text{Log} \frac{r_0}{a} \right) \frac{2}{t(\text{Log}\,\tau/t)^2}$$

where r_0 is the initial distance between the two radicals, a their distance at the first encounter, $a < r_0$, and τ the time between diffusive jumps.

The probability of reaction for a c product and T and S precursors may be calculated:

$$P_2^{(T)} = \frac{\lambda}{2} \frac{\mathcal{H}_m^2}{\omega^2} \text{Log} \frac{r_0}{a} \bigg/ \text{Log}\,[(\omega\tau)^{-1/2}]$$

$$P_2^{(S)} = \lambda \left(1 - \frac{\mathcal{H}_m^2}{2\omega^2} \text{Log} \frac{r_0}{a} \right) \bigg/ \text{Log}\,[(\omega\tau)^{-1/2}]$$

where the variables have their usual meaning as in 2.3. The difference compared to the usual case is the logarithmic dependence on the magnetic field when $\omega \sim \mathcal{H}_m$. A difference appears, due to the temperature dependence of τ. The thickness of the sample is calculated and found to be of the order of 20 Å. Such an experiment is possible.

Following the same idea DEUTCH [196] has outlined theoretically the effect of hydrodynamic on CIDNP spectra.

3. Isotopic Substitutions

It is well known that isotopic substitution plays an important role in NMR spectroscopy [1e, 1f]. Generally this substitution is used to obtain NMR chemical shifts of atoms in low natural abundance such as ^{13}C or ^{17}O, or to measure spin-spin coupling constants of equivalent nuclei.

LAWLER and EVANS [102q] have pointed out the value of deuterium labelling in the study of CIDNP. With D atoms, three characteristics are modified: spin multiplicities (D atoms have 1 as total spin momentum), the gyromagnetic ratio is 6.514 times less than the proton, and the mass is twice that of protons. As dipole-dipole interactions are proportional to the gyromagnetic ratio, interaction with D atoms is small. For instance, the hyperfine coupling constant A is divided by 6.514.

This important term may change the CIDNP spectra. LAWLER et al. [102q] give Eq. (28) for the case of a radical pair with a proton and a deuteron:

$$P_m^S(H) = 1 - \frac{3/8\, A_H^2\, \tau_c^2}{1 + A_H^2\, \tau_c^2},$$

$$P_m^S(D) = 1 - \frac{4/9 \times 3/2\, A_D^2\, \tau_c^2}{1 + 3/2\, A_D^2\, \tau_c^2}.$$

We must mention that relaxation times are shorter for D than for H because D has a quadrupole moment.

To our knowledge, only KAPTEIN [81, 130] has performed experiments in this field. He has studied the thermal decomposition of acetyl peroxide in hexachloroacetone by means of deuteroacetyl peroxide,

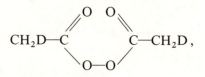

for which proton and deuterium NMR spectra are obtained with the CIDNP effect. For the ^2D spectrum, the multiplet effect observed on CH_2DCl and CH_2D—CH_2D is more pronounced than the net effect with the ^1H spectrum. This can be explained theoretically.

Equation (38) gives intensities for ^1H proportional to:

$$\tfrac{1}{2} A_H (\Delta g\, \beta_e\, H_0 + A_D M_D)$$

and for ^2D

$$\tfrac{1}{2} A_D (\Delta g\, \beta_e\, H_0 + A_H M_H)$$

where M_D has the values $-1, 0, +1$. As we have seen previously, the first term represents the net effect and the second represents the multiplet effect. Since

$A_H/A_D = 6.514$, the multiplet effect remains constant and the net effect is smaller for D than for H. The effect of isotopic substitution was recently discussed by KESSENIKH et al. [206].

LAWLER et al. [102q] also point out the importance of comparing the CIDNP effect on protons bound to ^{13}C and to ^{12}C. The first has a spin moment of $\frac{1}{2}$, the second has no spin. Some interactions seen in the first case disappear in the second. This is a way of testing the different contributions to the CIDNP effect. A similar situation may be found for ^{18}O, ^{17}O or ^{33}S, ^{32}S.

Such an interesting feature has been used by KAPTEIN [81, 130] who studied the thermal decomposition of acetyl peroxide (AP) in hexachlorobutadiene. Two experiments were performed under the same conditions, one with 55% ^{13}C-enriched AP, one with a mixture of 50% of natural AP and 50% of ^{13}C- enriched AP. This experiment makes it clear that not all relaxation processes in free radicals, such as the cross-relaxation mechanism in the Overhauser effect, can explain CIDNP. Only the radical-pair theory is in agreement with results of this experiment.

Chapter III. Applications to the Study of Chemical Reactions and Magnetic Properties

I. Applications to the Study of Chemical Reactions

For chemical reactions involving free radicals, CIDNP effects provide useful information on radical processes. These effects are caused by magnetic interactions in transient radical pairs and are influenced by the mode of formation and destruction of the pairs from which polarized products can be observed.

a) Radical-pair formation: the polarization is affected by the mode of formation of the pair: (i) singlet precursor; (ii) triplet precursor or pair generated by the encounter of independently generated free radicals.

The radical pairs generated from common precursors are said to be *correlated* or *geminate*, while the others are called *uncorrelated*, or *random-phase*, or *diffusive pairs*.

b) Radical-pair destruction: the pathway of product formation affects the nature of the CIDNP phenomenon: (i) radical-pair product formed by combination or dismutation (or disproportionation): cage product; (ii) subsequent reactions of polarized radicals leaving the pair: combination with other radicals or metathetical reactions with other molecules: escape product.

In the following scheme ($*$) indicates the change of phase of polarization:

$$
\begin{array}{c}
{}^1M \\
{}^3M \\
R. + R'.
\end{array}
\xrightarrow{(*)} \boxed{R. \ .R'} \xrightarrow{(*)}
\begin{array}{l}
\xrightarrow{\text{reaction}} \underline{P} \text{ (combination or dismutation)} \\
\xrightarrow{\text{diffusion}} \underline{R}. + \underline{R'}. \longrightarrow \text{subsequent reactions of} \\
\hspace{2.4cm}\Big\downarrow \text{relaxation} \hspace{1.3cm} \text{polarized radicals} \\
\hspace{2.3cm} R. + R'.
\end{array}
$$

Polarized radicals leaving the pair: $\underline{R}.$ and $\underline{R'}.$ can give three diffusive pairs: $\boxed{R. \ .R}$ $\boxed{R. \ .R'}$ $\boxed{R'. \ .R'}$. The polarization observed in the products of these pairs may be due in part to the former pair if the radicals $R.$ and $R'.$ have not recovered their Boltzmann equilibrium for nuclear spins (see 3.1.2.: "*memory effect*").

CIDNP effects provide evidence for radical intermediates. They can be observed when radicals are present either too briefly or in concentrations too low to be detectable by other methods. CIDNP effects must be use in combination with other techniques to elucidate the mechanism of reactions because minor radical reaction pathways can give rise to an important polarization. The observation of a CIDNP effect in a reaction cannot be taken as exclusive evidence for a radical mechanism.

During reactions where the products are chemically identical to the reactants, CIDNP effects demonstrate that a reaction is occurring.

Generally the polarization concerns the nuclei which are in proximity to the electrons of the radical pair and can give information as to the nature of the radical intermediates in the reaction.

This phenomenon can be observed for all nuclei possessing a nuclear spin. The majority of published examples deal with proton polarization, but CIDNP effects have been reported for ^{13}C, ^{19}F, ^{2}H, ^{31}P and ^{15}N.

CIDNP can be considered as a new and powerful research tool for kinetic studies and for investigating the mechanism of free-radical reactions.

KAPTEIN [81] (p. 13) has emphasized that the tiny magnetic interactions between electrons and nuclei ($\simeq 10^{-5}$ kcal/mole) are not capable of influencing the course of chemical reactions. Moreover, even when the reaction occurs to a small extent through radical-pair mechanism with nuclear-spin selection (say, for several %), it produces strong polarization enhancements because of the small difference of population at thermal equilibrium.

CIDNP effects have been reported for many reactions [76, 83b, 102, 143, 148, 175, 176c–176h, 213a] and after a rapid survey of the reaction types studied, we shall give some examples of chemical reactions where these effects provide useful information.

1. Reaction Types Studied by CIDNP

a) Thermal or photochemical decompositions of peroxides: [5, 6, 10, 11, 21, 22, 24, 40, 44, 48, 50, 66, 67, 68, 74, 80, 81, 91, 92, 93b, 102j, 102l, 105, 111, 116, 127, 130, 158, 160, 162, 174, 176d, 176e, 178, 179] (see 3.1.2.).

b) Reductions of diazonium salts with $NaBH_4$, NaOH etc.: [35, 39, 84, 102m]. ^{13}C-polarization using the pulse Fourier transform technique has been reported for these reactions [121].

c) Rearrangements: [28, 31, 33, 34, 36–38, 51, 54, 57, 60, 62–65, 71, 94, 98, 100, 102b, 102c, 102d, 124, 147, 176h, 181, 185, 191, 192, 213b] (see 3.1.5.).

d) Reactions of carbenes: [16, 43, 44, 102i, 125, 140, 176c]. CIDNP technique has been applied to determine the spin multiplicity of carbenes involved in abstraction-recombination reactions:

$$\begin{matrix} R \\ \diagdown \\ \quad C: + R''X \rightarrow \boxed{\begin{matrix} R \\ | \\ R'-C..R'' \\ | \\ X \end{matrix}} \rightarrow \begin{matrix} R \\ | \\ R'-C-R'' \\ | \\ X \end{matrix} . \\ \diagup \\ R' \end{matrix}$$

A similar result is obtained on nitrene [166].

e) Free-radical attacks on aromatic molecules: [88, 102e, 110]. During these reactions, CIDNP has given evidence as to the presence of an intermediate cyclohexadienyl-type radical:

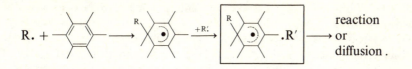

The radical pair is produced in a singlet state [88, 102e] or from independently generated free radicals [110].

Additions of free radicals (1-cyano-1-methylethyl radicals) to nitrones, nitrosobenzene [52] and 9-bromoanthracene [137] (see 3.1.4.).

f) Photochemical reactions of aldehydes [42, 45, 58, 83b, 131, 161, 176c, 193, 202, 204, 214], ketones [13, 17, 42, 45, 77, 81, 83a, 102n, 102o, 122, 163, 176c, 197], phenanthraquinone [95, 96, 97, 141, 142] and of quinones [167, 169, 176c, 194]. Photooxidation of phenol [132] and tyrosine [150] in water and of hydrosol suspensions of Congo Red [173]. Photochemical cleavage of carbon–sulfur bond [153].

g) Azo-compound reactions [176c]. The first examples of CIDNP effects were reported during azo-compound decompositions [10, 21]. These reactions have given evidence as to the validity of the radical-pair theory [19, 43, 44] (see 3.1.3). CIDNP has been reported for decomposition of diazo compounds [93a, 182], unsymmetrical azo compounds [119] and acid-catalyzed decomposition of 2-tetrazenes [145]; sigmatropic rearrangements of diazenes [100], photochemical decomposition of methyl diazoacetate [59]; free-radical reactions of 1,3,5-triarylpentazadienes (1,3 migration of arylazo groups) [61], and of 1-tetrazenes (decomposition) [102k]. For unsymmetrical azo compounds the loss of nitrogen may occur either by concerted cleavage of both C–N bonds or by a stepwise process giving rise to an intermediate diazenyl radical:

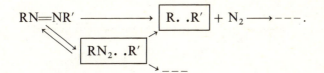

Because of the differences expected in the magnetic properties of the initial radical pairs associated with the two mechanisms, CIDNP can provide information which helps to distinguish between them [119].

h) Biradical reactions: dimerization of trimethylenemethane derivatives [82, 101, 102f], photolysis of alkyl O-benzoyl benzoates [102f], of cyclic ketones [172] and photochemical decomposition of cyclohexane diperoxide [78, 81].

In the radical-pair theory, in order to observe CIDNP signals in the products, for $S - T_0$ transitions, spin-independent competitive processes are required. They are provided by the diffusion of the components from the radical pair (see Chapt. 2). In the case of a biradical radical pair, such competitive processes do not exist if the reactions are intramolecular cyclization or disproportionation, and no

CIDNP can be expected with $S - T_0$ transitions ($\Delta m = 0$). With $S - T_{\pm 1}$ transitions ($\Delta m = \pm 1$) it is possible to observe CIDNP in biradicals under the following conditions [102f]: i) the average exchange interaction should be of the order of the Zeeman splitting ($J \simeq \frac{1}{2} g \beta_e H_0$, to allow $S - T_{\pm 1}$ transitions). By varying the magnetic field, CLOSS and DOUBLEDAY [172] have determined J in biradicals resulting from α cleavage of ketones (There is a maximum in the CIDNP effect intensity vs. field curve). ii) the biradical must be produced in the triplet state ("net" transitions are required). The only case observed from triplet precursor is given by CLOSS et al. [152].

However, CIDNP effects arising from spin selection have been observed in biradical products from singlet precursors [78, 81], the competing reaction, independent of the electronic multiplicity of the biradical, being a transfer reaction with the solvent.

i) Reactions of organometallic compounds with alkyl and aromatic halides [8, 12, 14, 15, 24–27, 29, 30, 32, 102p, 176g, 208, 215], reactions of Grignard reagents with alkyl halides [72, 108, 165, 184] or with diazonium salts [190], iodine transfer reactions [87], reaction of alkyl iodides with sodium mirrors (Wurtz reaction) [55b] and of alkyl halides with sodium naphthalene [55a, 99, 176g]. Grignard reagents with alkyl halides are quoted to students as an example of CIDNP [164].

j) Decomposition reactions of dimethyl perdiglutarate [106], di-t-butyl-peroxalate [205], O–O'-diacyl-4-hydroxyaminoquinoline 1-oxide [86], N,N'-diacetyl-1,1', 4,4'-tetrahydrobipyridine [102h], N,N'-diacetyl-tetrahydrodipyridyl-4,4' [154], N,N'-diethoxycarbonyl-4,4', N,N'-tetrahydro-4,4'-dipyridyl [151] N-nitrosohydroxylamines [70], organomercury derivatives of tin [t-BuHgSn(Me)$_3$ and t-BuHgSn(Et)$_3$] [120] and of phenyldiimine [138]. Reaction of 4-picoline N-oxide with acetic anhydride [146, 177, 211] and oxydation of dialkylsulphides with nitric acid [183].

k) Radical ions: CIDNP effects observed during the cleavage of thymine dimers sensitized by quinones have been attributed to the pair of radical ions [115]

$$\boxed{A^- . . B^+}$$

$A = $ 2-anthraquinone sulfonate

$B = $ 1,3-dimethylthymine

Nucleophilic and electrophilic substitutions [187, 207]: intermediate radical particles resulting from processes of electron transfer have been detected during these reactions.

For reactions of radical ions, CIDNP can be an useful mechanistic tool. The reactions of alkyl halides with sodium naphthalene which have been described by GARST [176h] illustrate the possibility of investigation by CIDNP in the field of radical-ion chemistry.

l) Other nuclei:
^{13}C: [66, 81, 102j, 105, 121, 127, 130, 176d, 177, 182, 185, 186, 188, 211] (see 3.1.6.)
^{19}F: [53, 85, 88, 102e, 107, 118, 155a, 155b, 159, 198, 209] (see 3.1.7.)
^{31}P: [69, 199–201] (see 3.1.8)

[15]N: [*102j, 186, 188*] (see 3.1.8.)
[2]H: [*81, 130*] (see 2.6.3.).

2. Thermal or Photochemical Decompositions of Peroxides

These decomposition reactions are very suitable for the observation of CIDNP effects, and for this reason have been the subject of many studies.

One of the first examples of CIDNP, the benzene NMR emission during the thermal decomposition of dibenzoyl peroxide in cyclohexanone, was reported by BARGON and FISCHER [5]. Such effects of nuclear polarization were observed during the decomposition of various peroxides [*6, 10, 11, 21, 22, 24, 67*] and were explained by electron–nucleus cross-relaxations in intermediate free radicals [*6, 21*].

These studies were repeated in order to test the new theory of the radical pair. The dependence of product distribution and CIDNP patterns on the concentrations of radical-trapping agents during the thermal decomposition of di-(4-chlorobenzoyl)-peroxide have been reported by BLANK and FISCHER [*74, 102l*]. They demonstrate, in agreement with KAPTEIN et al. [*50*], that the CIDNP effects are caused by $S - T_0$ transitions in transient benzoyloxy/phenyl radical pairs:

$$\boxed{\varnothing\text{—}CO_2.\ .\varnothing}$$

The photolysis of dibenzoyl peroxide in binary solvents $(RX/R'X')$ has been reported by LEHNIG and FISCHER [*48*]. This radical reaction is represented by:

$$\varnothing\text{—}\underset{\underset{O}{\|}}{C}\text{—}O\text{—}O\text{—}\underset{\underset{O}{\|}}{C}\text{—}\varnothing \longrightarrow 2\varnothing\text{—}CO_2. \longrightarrow 2\varnothing. + 2CO_2 \qquad (1)$$

$$\varnothing\text{—}\underset{\underset{O}{\|}}{C}\text{—}O\text{—}O\text{—}\underset{\underset{O}{\|}}{C}\text{—}\varnothing \longrightarrow \varnothing\text{—}CO_2\text{—}\varnothing + CO_2 \qquad (2)$$

$$\varnothing. + RX \longrightarrow \varnothing X + R. \qquad (3)$$

$$\varnothing. + R'X' \longrightarrow \varnothing X' + R'. \qquad (4)$$

$$R. + R'. \longrightarrow R - R' \qquad (5)$$

$$R. + R. \longrightarrow R - R \qquad (6)$$

$$R'. + R'. \longrightarrow R' - R' \qquad (7)$$

$$R. + R'X' \longrightarrow RX' + R'. \qquad (8)$$

$$R'. + RX \longrightarrow R'X + R. \qquad (9)$$

The decomposition products of peroxide: $\emptyset X$ and $\emptyset X'$ generally show emission, the radical pair $\boxed{\emptyset\text{—}CO_2.\ .\emptyset}$ being generated from a singlet state during the thermal decomposition or photolysis of the dibenzoyl peroxide, or enhanced absorption during photodecomposition with triplet sensitizers [92] (opposite effect for $\emptyset\text{—}CO_2\text{—}\emptyset$).

The benzene emission can be explained by *Kaptein's rules*:

$$\boxed{\emptyset.\ .CO_2\text{—}\emptyset}$$

$$\Gamma_{np} = \mu.\varepsilon.\Delta g.A$$

μ: $(-)$, pair generated by a precursor in singlet state.
ε: $(-)$, transfer reaction product, polarized phenyl radical having escaped the radical pair.
Δg: $(-)$, $g(\emptyset.) < g(\emptyset\text{—}CO_2.)$.
A: $(+)$, the hyperfine coupling constant for the *ortho* protons responsible for emission is positive [3].

$$\Gamma_{np} = (-)(-)(-)(+) = (-) : \text{emission}\,(E).$$

For phenyl benzoate, the same protons give enhanced absorption, in this case $\varepsilon = (+)$, for a cage combination product.

The CIDNP patterns of the products formed during reactions (5) to (9) are independent of the peroxide used. These CIDNP effects may be attributed to $T_0 - S$ transitions in pairs:

$$\boxed{R.\ .R'}\ , \quad \boxed{R.\ .R} \quad \text{and} \quad \boxed{R'.\ .R'}$$

formed by diffusive encounters of radicals $R.$ and $R'..$ In these examples the symmetrical products $(R - R, R' - R')$ have single line NMR spectra and multiplet polarizations are not observable. The observed CIDNP effects (net polarization) are attributed to transitions in the unsymmetrical radical pair $\boxed{R - R'}$ and

---

[3] With di-(4-chlorobenzoyl)-peroxide, in the presence of iodine, the protons of 1-iodo-4-chlorobenzene are not magnetically equivalent, and no CIDNP effect is observed for the *meta* protons [74].

transferred to other products according to the following scheme[4]:

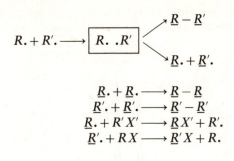

$$R. + R. \longrightarrow \underline{R} - \underline{R}$$
$$R'. + R'. \longrightarrow \underline{R'} - \underline{R'}$$
$$\underline{R}. + R'X' \longrightarrow \underline{R}X' + R'.$$
$$\underline{R'}. + RX \longrightarrow \underline{R'}X + R.$$

The polarizations of symmetrical coupling $(R - R,\ R' - R')$ and transfer $(RX',\ R'X)$ products have opposite signs to those of corresponding groups of the unsymmetrical products $(R - R')$. The polarizations of $R.$ and $R'.$ built up in the radical pair $\boxed{R.\ .R'}$ decay by nuclear relaxation in the free radicals. This scheme explains why the amplitude of the CIDNP effects are generally much lower for the symmetrical coupling and transfer products than for the unsymmetrical coupling products. These results are shown in the Fig. 3.1, for the products $R - R', R - R$ and $R' - R'$ from the photolysis of dibenzoyl peroxide in the binary solvent CH_2Cl_2/CH_3COCH_3.

The polarization of $C\underline{H}Cl_2 - R'$ depends on the Δg value of the radicals $.CHCl_2$ and $R'.$. Thus an emission is observed for $C\underline{H}Cl_2$—CHClCOOH and an enhanced absorption for $C\underline{H}Cl_2$—CCl$_3$, and it is deduced that:

$$g'(.CHClCOOH) < g(.CHCl_2) < g'(.CCl_3)$$

$$2.0067 < g(.CHCl_2) < 2.0091\ .$$

On the basis of Kaptein's rules the authors have demonstrated that the polarization observed for $C\underline{H}Cl_2 - R'$ is compatible with a negative hyperfine coupling constant in the radical $.CHCl_2$. From comparisons of calculated and observed CIDNP effects the g factor and the proton hyperfine coupling constant of $.CHCl_2$ have been determined (see 3.2.2 and 3.2.3).

The enhancements of the symmetrical and transfer products have been quantitatively discussed. The polarization is produced in unsymmetrical pairs and decay by nuclear relaxation in the free radicals $R.$ and $R'.$; for the radical $.C\underline{H}Cl_2$ the relaxation time T_R has been evaluated: $T_R \simeq 4.5\ 10^{-4}$ sec (see 3.2.5a).

The thermolysis of benzoyl peroxide in solutions containing alkyl iodides has been reported by COOPER et al. [111b]. During this reaction, the CIDNP

---

[4] See further the "memory effect" reported by KAPTEIN et al. [81, 130].

spectra of alkyl benzoates have been observed and must result from interactions in benzoyloxy/alkyl geminate pairs (and not diffusive pairs). These pairs may be formed by abstraction of iodine from alkyl iodide by a geminate partner of the benzoyloxy radical. This abstraction proceeds with spin conservation and the spin multiplicity of the new radical pair is the same as that of its predecessor pair.

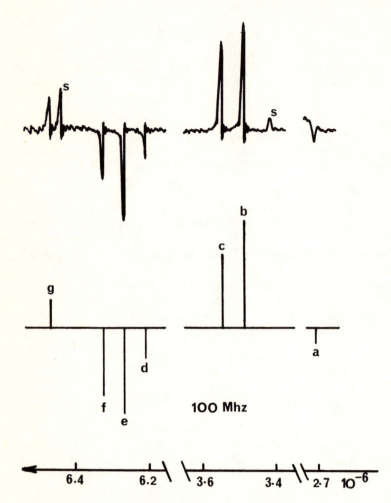

Fig. 3.1. Photolysis of dibenzoyl peroxide in a binary solvent (CH_2Cl_2/CH_3COCH_3)
$R. = .CHCl_2$ $R'. = .CH_2COCH_3$
1) $R–R'$ $C\underline{H}Cl_2–CH_2COCH_3$ $\delta = 6.32\ 10^{-6}\ E\ (f)$
                                                         $\delta = 6.26\ 10^{-6}\ E\ (e)$
                                                         $\delta = 6.20\ 10^{-6}\ E\ (d)$
              $CHCl_2–C\underline{H}_2COCH_3$ $\delta = 3.53\ 10^{-6}\ A\ (c)$
                                                         $\delta = 3.47\ 10^{-6}\ A\ (b)$
2) $R–R$ $C\underline{H}Cl_2–C\underline{H}Cl_2$ $\delta = 6.47\ 10^{-6}\ A\ (g)$
3) $R'–R'$ $CH_3COC\underline{H}_2–C\underline{H}_2COCH_3$ $\delta = 2.68\ 10^{-6}\ E\ (a)$
(From Lehnig and Fischer [48])

This type of radical trapping is called "*pair substitution*":

$$(\text{ØCO}_2)_2 \rightarrow \boxed{\text{ØCO}_2. \ .\text{Ø}}^{\text{S}} \xrightarrow{+\text{RI}} \boxed{\text{ØCO}_2. \ .R}^{\text{S}} \begin{cases} \nearrow \text{ØCO}_2\text{H} + R(-H) \\ \\ \searrow \text{ØCO}_2\text{R} \end{cases}$$

CIDNP spectra of $\text{ØCO}_2\text{H}$, $R(-H)$ and $\text{ØCO}_2\text{R}$ observed during the thermolysis of benzoyl alkyl peroxide $\left(\text{geminate radical pair:} \ \boxed{\text{ØCO}_2. \ .R}\right)$ are entirely analogous to those resulting from pair substitution. A similar effect has been reported by KAPTEIN et al. [80] during the thermal decomposition of isobutyryl peroxide with bromoacetone: the sign of polarization of chloroform changes with increasing bromoacetone concentration (emission → enhanced absorption):

$$(\text{CCl}_3\text{Br}) < 0.1\,\text{M}:$$

$$R. + .\text{CCl}_3 \rightleftharpoons \boxed{R. \ .\text{CCl}_3}^{\text{F}} \longrightarrow \text{C}\underline{\text{H}}\text{Cl}_3 + R(-H)$$

$$(\text{CCl}_3\text{Br}) > 0.1\,\text{M}:$$

$$(\text{RCO}_2)_2 \longrightarrow \boxed{R. \ .R}^{\text{S}} \xrightarrow{+\text{CCl}_3\text{Br}} \boxed{R. \ .\text{CCl}_3}^{\text{S}} \longrightarrow \text{C}\underline{\text{H}}\text{Cl}_3 + R(-H)$$

The thermal decomposition of acetyl peroxide in hexachloroacetone [81, 130] gives rise to the following effects: emission for ethane ($\delta = 0.83$ ppm) and methyl acetate (OCH_3, $\delta = 3.54$ ppm), and enhanced absorption for methyl chloride ($\delta = 2.94$ ppm) and methane ($\delta = 0.18$ ppm) (Fig. 3.2). The following scheme has been proposed by several authors [81, 111a, 130]:

$$(\text{CH}_3\text{CO}_2)_2 \rightleftharpoons \boxed{\text{CH}_3\text{CO}_2. \ .\text{CO}_2\text{CH}_3}$$

$$\downarrow$$

$$\text{CO}_2 + \boxed{\text{CH}_3. \ .\text{CO}_2\text{CH}_3} \longrightarrow \text{CH}_3\text{CO}_2\text{CH}_3$$

$$\downarrow$$

$$\text{CO}_2 + \boxed{\text{CH}_3. \ .\text{CH}_3} \begin{cases} \longrightarrow \text{C}_2\text{H}_6 \\ \searrow \text{CH}_3. + \text{CH}_3. \end{cases}$$

$$\text{CH}_3. + \text{RX} \longrightarrow \text{CH}_3\text{X} + R.$$

During this thermal decomposition, the ethane emission cannot originate from the pair of equivalent methyl radicals ($\Delta g = 0$). The polarization of methyl acetate and that of ethane have a common origin: $S - T_0$ mixing in the methyl/acetoxy radical pair. This phenomenon is called a "*memory effect*" by KAPTEIN [*81, 130*], because polarization can result from all pairs preceding the pair from which the product is actually formed.

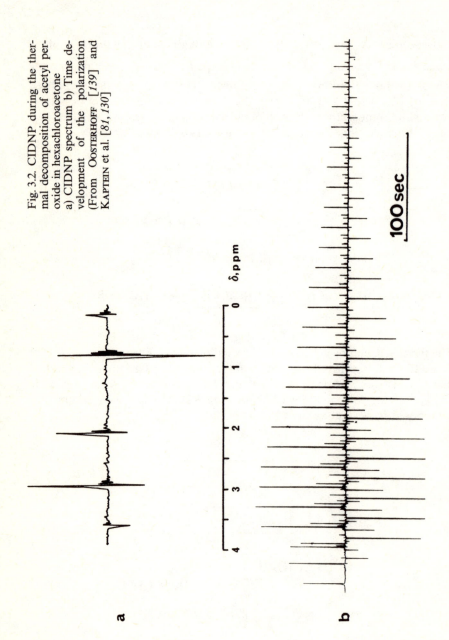

Fig. 3.2. CIDNP during the thermal decomposition of acetyl peroxide in hexachloroacetone a) CIDNP spectrum b) Time development of the polarization (From OOSTERHOFF [*139*] and KAPTEIN et al. [*81, 130*])

The CIDNP spectra obtained during the decomposition of propionyl and all of the higher n-alkyl peroxides differ markedly from those observed during the decomposition of acetyl peroxide in that they exhibit only multiplet polarization [111a]. For propionyl peroxide, the "memory effect" (net polarization) is not observed. The butane formed from the geminate pair of ethyl radicals shows a multiplet effect:

$$(C_2H_5CO_2)_2 \longrightarrow \boxed{C_2H_5CO_2 . .C_2H_5} \longrightarrow \boxed{C_2H_5. .C_2H_5} \longrightarrow C_4H_{10} \quad (E/A)$$

$$\downarrow$$

$$C_2H_5. + RX \longrightarrow C_2H_5X + R.$$

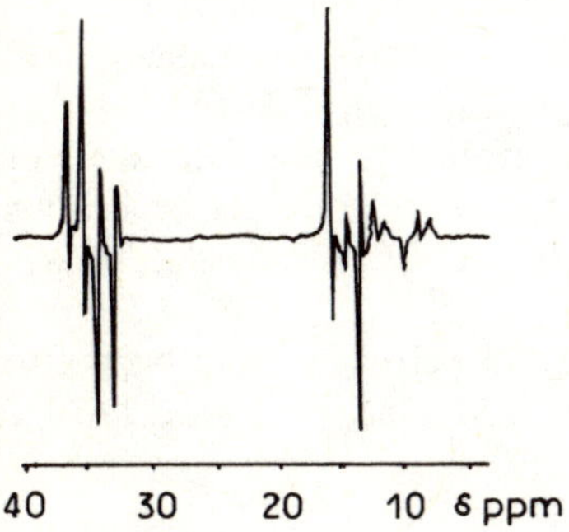

Fig. 3.3. CIDNP spectrum taken during the decomposition of propionyl peroxide in hexachloroacetone CIDNP effect for ethyl chloride: CH_3CH_2Cl (CH_2): $\delta = 3.52$ ppm (CH_3): $\delta = 1.47$ ppm (From KAPTEIN [11,81])

During the thermal decomposition of propionyl peroxide in hexachloroacetone [11,81], the *multiplet polarization* of ethyl chloride (Fig. 3.3) can be explained by KAPTEIN's *rules* [81, 135]:

$$\boxed{CH_3CH_2 . .CH_2CH_3}$$

$$\Gamma_{me} = \mu . \varepsilon . A_i . A_j . J_{ij} . \sigma_{ij}$$

μ: $(-)$, pair generated by a precursor in a singlet state.
ε: $(-)$, transfer reaction product, polarized ethyl radical having escaped the radical pair.
A_i: $(-)$ ⎰in the ethyl radical α protons (CH_2) have a negative hyperfine coupling
A_j: $(+)$ ⎱constant $(-22.4\,G)$ and β protons (CH_3) a positive value $(+26.9\,G)$.
J_{ij}: $(+)$, vicinal coupling constant being $+7\,Hz$ for ethyl chloride.
σ_{ij}: $(+)$, nuclei i and j belong to the same radical.

$$\Gamma_{me}(CH_2) = (-)(-)(-)(+)(+)(+) = (-) : A/E .$$

$$\Gamma_{me}(CH_3) = (-)(-)(+)(-)(+)(+) = (-) : A/E .$$

There is a small net polarization (CH_3: E, CH_2: A) superimposed on the multiplet effects in ethyl chloride and the radical pair responsible for this effect may be the pentachloroacetonyl/ethyl radical pair [81, 176d].

Characteristics of the CKO theory are that net polarization (E or A) arises only from pairs of non-equivalent radicals (Δg effect) and that pairs of equivalent radicals ($\Delta g = 0$) can only give multiplet effects. On this basis KAPTEIN [81, 176d] distinguishes two classes of peroxides.

Class (a): peroxides which give rise to net polarization due to the pair $\boxed{RCO_2. \,.R}$: acetyl peroxide, benzoyl peroxide, etc.

Class (b): peroxides which only exhibit multiplet effects due to the pair $\boxed{R. \,.R}$ (net effects, in this case, can be due to solvent-derived radicals): propionyl peroxide, butyryl peroxide, etc. In this case the acyloxy radical ($RCO_2.$) decarboxylates too rapidly and cannot give rise to observable net effects.

The phenylacetyl peroxide belongs to class (b); WALLING and LEPLEY have reported the thermal decomposition of this peroxide in different solvents (CCl_3Br, CCl_3SO_2Cl, etc.) [91, 116]. They have evaluated rate constants for benzyl radical-substrate reactions. (Polarization for benzyl radicals occurs via sorting accompanying diffusive encounters of benzyl and $CCl_3.$ radicals: "*radical flux*" model.)

The thermal or photochemical decompositions of peroxides have been also investigated by [13]C—CIDNP [66, 81, 102j, 105, 127, 130, 176d, 179] (see 3.1.6) and [2]H—CIDNP [81, 130].

3. Photochemical Reactions

In 1968 COCIVERA [13] reported the observation of nuclear polarization (emission) in the proton NMR spectrum of anthraquinone induced by optical excitation of this molecule to its lowest electronic triplet state. This phenomenon was interpreted by the Overhauser effect.

In 1969 CLOSS and CLOSS reported the observation of CIDNP during reactions of photochemically generated triplet diphenylmethylene [16] and photoreductions of benzophenone [17]. From the radical-pair theory, the phase of polarization ($A - E$; $A/E - E/A$) depends on the multiplicity of the precursor (triplet or singlet). This effect was demonstrated by CLOSS and TRIFUNAC [19]. Photochemical reactions showing CIDNP effects were interpreted with the CKO theory in 1970 by CLOSS et al. [42–45], by KAPTEIN et al. [50] and confirmed the validity of the radical-pair theory.

In photochemistry, CIDNP can contribute to the determination of the spin multiplicity of the radical pair precursor [175].

In 1969 CLOSS and TRIFUNAC [19] demonstrated that the multiplet effect changes phase ($A/E - E/A$) with the multiplicity of the precursor for 1,1,2-triphenylethane.

Diphenyldiazomethane (I) decomposes photochemically to diphenylmethylene (II) which has a triplet ground state [16]. In toluene the benzhydryl/benzyl radical pair is formed from a T precursor:

$$\emptyset_2CN_2 \xrightarrow[-N_2]{h\nu} \emptyset_2C: \xrightarrow{+\emptyset CH_3} \boxed{\emptyset_2CH. \ .CH_2\emptyset}^T$$

$$\text{(I)} \qquad\qquad \text{(II)} \qquad\qquad\qquad \text{(III)}$$

Photochemical (or thermal) decomposition of the corresponding azo compound (IV) gives the same radical pair, except that its precursor is now a singlet:

$$\emptyset_2CHN{=}NCH_2\emptyset \xrightarrow[-N_2]{h\nu} \boxed{\emptyset_2CH. \ .CH_2\emptyset}^S$$

$$\text{(IV)} \qquad\qquad\qquad \text{(V)}$$

The two radical pairs (III) and (V) differing in the multiplicity of the precursor give the same recombination product: 1,1,2-triphenylethane:

$$\emptyset_2C\underline{H}C\underline{H}_2\emptyset$$

$$\text{(VI)}$$

In Fig. 3.4, the NMR spectra of (VI) (A_2B system of benzylic protons, lines 1 to 8) taken during these two reactions show opposite polarizations.

KAPTEIN et al. [50], in a report on the photolysis of some peroxides in the presence of photosensitizers, have underlined a difficulty in the determination

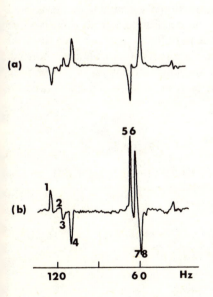

Fig. 3.4. Polarized NMR spectra of 1,1,2-triphenylethane (a) S Precursor: thermal decomposition of $\emptyset_2CHCN{=}NCH_2\emptyset$ (b) T precursor: photolysis of diphenyldiazomethane in toluene (From CLOSS and TRIFUNAC [19])

of the precursor multiplicity. The theory predicts that reaction products of diffusive free radicals are polarized like T-pair products, so that when T-pair polarization is observed, two explanations are possible:

   i) cage recombination of a T pair,

   ii) combination from secondary encounters of free radicals.

For a radical pair generated from free radicals, the polarization is of smaller magnitude than for the T precursor, but quantitative CIDNP theories are not yet able to distinguish between these two cases.

Ketones which undergo the Norrish type-I photocleavage have been found to lead to strong CIDNP effects. During the photolysis of benzoin methyl ether DOMINH [102n] observed a strong polarization in the spectrum of the starting material. This suggests a cleavage-recombination process in which the methine proton appears as an emission signal:

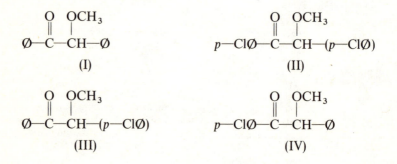

This polarization is consistent with the recombination of a geminate pair generated from the triplet precursor [process (1)] or the combination from diffusive free radicals [process (2)]. The question of geminate or diffusive pairs has been resolved by a CIDNP experiment in which two independent pairs of radicals are produced. Irradiation of a mixture of (I) and (II) gives four emission signals of comparable intensity due to the methine protons of (I) to (IV), as expected for a random combination of the radicals [process (2)].

$$\underset{\text{(I)}}{\overset{\displaystyle O\quad OCH_3}{\Phi-\overset{\|}{C}-\overset{|}{C}H-\Phi}}\qquad\qquad\underset{\text{(II)}}{\overset{\displaystyle O\quad OCH_3}{p-Cl\Phi-\overset{\|}{C}-\overset{|}{C}H-(p-Cl\Phi)}}$$

$$\underset{\text{(III)}}{\overset{\displaystyle O\quad OCH_3}{\Phi-\overset{\|}{C}-\overset{|}{C}H-(p-Cl\Phi)}}\qquad\qquad\underset{\text{(IV)}}{\overset{\displaystyle O\quad OCH_3}{p-Cl\Phi-\overset{\|}{C}-\overset{|}{C}H-\Phi}}$$

In this case the CIDNP effect cannot be used for the determination of the precursor multiplicity.

During the photochemical (or thermal) decomposition of dibenzoyl peroxide the benzene which is formed gives an emission spectrum. FAHRENHOLTZ and TROZZOLO [92] have reported examples of an enhanced absorption of benzene which occurs when certain triplet sensitizers are present:

$$\emptyset—CO_2—CO_2—\emptyset \begin{cases} hv + \text{singlet sensitizer} \rightarrow C_6H_6\ (E) \\ hv + \text{triplet sensitizer} \rightarrow C_6H_6\ (A). \end{cases}$$

They observed enhanced absorption only with sensitizers having a triplet energy of 59 kcal mole^{-1} or greater. Sensitization by excited singlet state of anthracene has been observed and it results in an increase in the intensity of the emission signal [50].

During photolysis of diisopropyl ketone, CIDNP spectra are consistent with the formation of an acyl/alkyl radical pair from the ketone triplet state. This radical pair results from a Norrish type-I split and CIDNP effects for this reaction have been observed for several ketones [77, 102n, 102o, 122].

In CDCl$_3$, DEN HOLLANDER et al. [77, 102o] have observed CIDNP effects which are compatible with the following scheme:

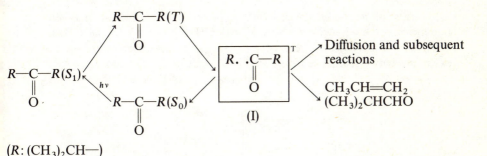

$(R: (CH_3)_2CH—)$

In CCl$_4$ the photolysis of diisopropyl ketone shows CIDNP effects, and the polarized spectrum entails the formation of a CCl$_3$./isopropyl radical pair from a singlet state precursor in contrast to the previous case (DOMINH has reported a similar effect for isomesityl oxide in CCl$_4$ [102n]). Two possible pathways are suggested:

The second mechanism is favored, for when CCl_4 is added it quenches the fluorescence of diisopropyl ketone. It has been noticed in the NMR spectrum that intersystem crossing is not suppressed completely, a small fraction of S_1 ketones apparently arrives in the T state and reacts as in $CDCl_3$. In this case, CIDNP effects from both S and T pairs derived from the same parent compound appear in the same spectrum.

Irradiation of a solution of propionaldehyde results in CIDNP which is exhibited in the NMR spectrum for several compounds [204]. The relative intensity of the polarized lines as well as the direction of the line due to the carbonyl hydrogen of propionaldehyde depends on the solvent. The solvent has an effect on the relative rates of the primary processes: in

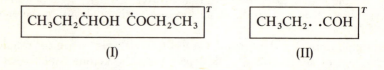

(I)   (II)

perfluoromethylcyclohexane, radical pair (I) predominates in determining the CIDNP for the carbonyl hydrogen; in deuteroacetonitrile, perdeuterobenzene, t-butyl alcohol or hexane, radical pair (II) predominates in determining the polarization.

During photochemical reactions of phenanthraquinone with o-substituted toluenes involving hydrogen abstraction, a CIDNP effect from the 1,2-adduct (II) has been observed [96]. The experimental results have been interpreted in terms of mixing between the electronic singlet (S) and triplet (T_0 and T_{+1}) states of the associated radical pair (I) (see general treatment of the CKO theory, 2.3):

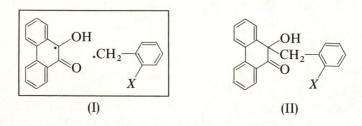

(I)   (II)

The exchange integral J has a positive value which is intermediate between the Zeeman energy and the hyperfine-coupling constant (see 3.2.4).

4. Applications of CIDNP to Chemical Kinetics

BUCHACHENKO and MARKARIAN have published a consideration of CIDNP kinetics and its relationship to chemical kinetics [133]. Quantitative studies of the mechanisms and kinetics of chemical reactions can be made by CIDNP.

In the general case of any reaction, the CIDNP kinetics are described by the following equation:

$$\frac{dI}{dt} = E\left(\frac{dI_0}{dt}\right) - \beta(I - I_0).$$

I is the nuclear magnetization (or NMR signal) of molecules considered, I_0 is their equilibrium magnetization; $\beta = T_1^{-1}$ is the rate constant of nuclear relaxation (T_1 = nuclear relaxation time, without considering cross-relaxation processes); E is the enhancement coefficient of nuclear polarization (ratio of nuclear polarization at the moment of its formation to the equilibrium polarization of this molecule).

Experimentally it is more convenient to consider the kinetic behaviour of the coefficient V:

$$V = \frac{I - I_{0\infty}}{I_{0\infty}}.$$

Its expression has been given for various kinetic reactions:
  a) first-order reaction:

$$A \xrightarrow{k} \underline{P}$$

  b) competitive reactions:

$$A \begin{cases} \xrightarrow{k_1} \underline{P}_1 \\ \xrightarrow{k_2} \underline{P}_2 \end{cases} \qquad A \begin{cases} \xrightarrow{k_1} \underline{P} \\ \xrightarrow{k_2} \underline{P} \end{cases}$$

  c) consecutive reactions:

$$A \xrightarrow{k_1} B \xrightarrow{k_2} \underline{P}$$

  d) bimolecular reaction:

$$A + B \xrightarrow{k} \underline{P}$$

e) zero-order reaction (photochemical initiation):

$$A \longrightarrow \underline{P}$$

($\underline{P}$ is the product molecule with polarized nuclei).

BUCHACHENKO et al. [93a, 102g, 133, 189] have underlined the applications of these kinetic equations to the determination of parameters concerning chemical kinetics and CIDNP: rate constants, activation energies, enhancement factors, etc.

For instance, during the reaction of aniline and isoamyl nitrite, two time-separated maxima of CIDNP for phenyl protons have been observed. The product reaction (benzene) is generated by two successively formed sources of phenyl radicals:

$$\xrightarrow{(1)} \varnothing. + P + R. \xrightarrow{(2)} \varnothing. + P + R'.$$
$$\downarrow \qquad\qquad\qquad \downarrow$$
$$\underline{\varnothing H} \qquad\qquad\qquad \underline{\varnothing H}$$

Processes (2) proceed approximately 20 times slower than processes (1). The kinetic parameters for this reaction have been determined by LEVITT et al. [93a, 189].

IWAMURA et al. [52] have reported an example in which CIDNP can be applied to the assignment of the NMR spectra. The thermal decomposition of α,α'-azo-bisisobutyronitrile (AIBN) in the presence of α, N-diphenylnitrone (I) proceeds through the initial attack of a 1-cyano-1-methylethyl radical to the carbon atom of the nitrone group to form the intermediate nitroxide (II), followed by the

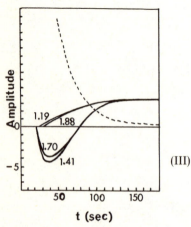

t (sec)

Fig. 3.5. Time-dependence of the amplitude of the four C-methyl signals of (III)

$$\overset{*}{C}Me_2CN \qquad\qquad (*) \quad \delta = 1.19–1.88 \text{ ppm}$$
$$|$$
$$(III) \quad \varnothing—CH—N—\varnothing$$
$$|$$
$$O—C\overset{**}{M}e_2CN \quad (**) \; \delta = 1.41–1.70 \text{ ppm}$$

Broken line: NMR absorption of AIBN at $\delta = 1.66$ ppm (From IWAMURA et al. [52])

addition of a second free radical to give the trisubstituted hydroxylamine (III):

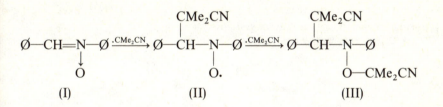

The addition product (III) shows four lines for the methyl protons $(\delta = 1.19\text{–}1.41\text{–}1.70\text{–}1.88)$, and polarization is only observed for two lines (net polarization, emission, $\delta = 1.41\text{–}1.70$, Fig. 3.5). The methyl signals which appear in emission can be assigned to the $-OCMe_2CN$ of the hydroxylamine (III)[5], the polarization originating from the recombination of radicals (II) and $.CMe_2CN$. This method can have an inverse application to demonstrate the chronology of free radical additions when the assignment of the spectra is unambiguous.

5. Rearrangements

The observation of the CIDNP effect during a rearrangement necessitates a radical process: homolytic dissociation, radical pair formation, and recombination of the radicals. CIDNP can be used as a tool to establish the mechanism of rearrangements [63, 102b, 102d, 176h, 213b].

The CIDNP effect was observed in 1969 by Lepley [31] during a benzyne addition–rearrangement reaction. During this reaction of benzyne adduct to N,N-dimethylbenzylamine, a multiplet effect (E/A) was observed for the quartet from the benzylic methine proton of the reaction product: N-methyl-N(α-phene-thyl) aniline (I). The following scheme in which a radical pair gives rise to the product (I) has been proposed:

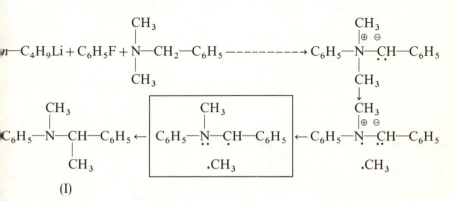

(I)

---

[5] Iwamura et al. have interpreted this example by the Overhauser effect in radicals. Their assignment of the NMR lines to the methyl protons may also be obtained from the radical-pair theory.

The radical pair which causes nuclear polarization is basically a specific type of representation for a Stevens rearrangement.

Since that study, the research and observation of the CIDNP effect have been reported for the following rearrangements:

(i) Wittig rearrangement [63, 102b]:

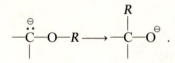

(ii) Meisenheimer rearrangement (Intramolecular rearrangement of a tertiary amine oxide) [57, 65, 102b]:

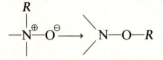

(iii) Sulfonium ylid-sulfide rearrangement [33, 60, 64, 94, 102b]:

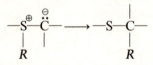

(iv) Stevens rearrangement [28, 31, 34, 54, 102b, 147]:

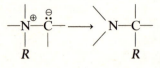

(v) Wawzoneck rearrangement [36, 37, 102d]:

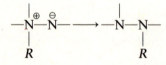

(vi) Martynoff rearrangement [98, 102d]:

$$X{=}N\overset{\oplus}{-}\overset{\ominus}{O}\longrightarrow X{=}N{-}O{-}R\,.$$
$$\underset{R}{|}$$

LEMAIRE and SUBRA [102c] have pointed out that many reactions, which at first were considered to occur by ionic processes, are shown to be radical reactions on the CIDNP basis. They propose another possibility: that the studied reactions are neither purely ionic, nor purely, radicalar, but may be described as a reso-nance between the two limiting forms. In the limiting pure ionic case no polariza-tion at all is possible.

For sigmatropic rearrangements in which two mechanisms are possible: concerted reaction (reaction proceeding with conservation of orbital symmetry) or radical dissociation–recombination mechanism (reaction proceeding through an intermediate radical pair), the observation of the CIDNP effect is evidence of the radical reaction [38, 51, 62, 100, 124, 185, 191, 192]. For example, the radical process for a 1,3-sigmatropic rearrangement is:

CIDNP has been detected in the Fries [181] and Claisen [195] photo-rearrangements. The photolysis of p-cresyl p-chlorobenzoate affords p-chloro-benzoyl p-cresol as the major product via aryloxy-aroyl radical pair derived from a singlet excited state precursor [181]. Experimental enhancement factors agree with calculated values, and it is unlikely that a concerted pathway is of major importance in this Fries photorearrangement.

However, the observation of nuclear polarization during a rearrangement is not sufficient in order to demonstrate the existence of a radical pair [192]. For example, it has been demonstrated previously that the rearrangement of benzyl toluene-p-sulphenate to benzyl p-tolyl sulphoxide:

$$C_6H_5{-}CH_2{-}O{-}S{-}C_6H_4{-}CH_3 \longrightarrow C_6H_5{-}CH_2\underset{\overset{\|}{O}}{-}S{-}C_6H_4{-}CH_3$$

does not involve a radical pair. The observation during this reaction of polarized NMR spectra with relatively large enhancements indicates that minor reaction pathways may give rise to observable signals [71].

6. CIDNP of ^{13}C

^{13}C-CIDNP effects have been reported for the decomposition of peroxides [66, 81, 102j, 105, 127, 130, 176d, 179] and for some other reactions (reduction of diazonium salts [121], reaction of 4-picoline N-oxide with acetic anhydride [177, 211], reactions of cyclohexadienone carbenes [182], 1,3 rearrangement of oxime thionocarbamates [185], thermal decomposition of diazoaminobenzene [102j, 186] (see 3.1.8.) and diazo coupling reactions [188]).

SCHULMAN et al. [127] have observed a polarized ^{13}C spectrum during the thermal decomposition of benzoyl peroxide in tetrachloroethylene. This spectrum exhibits several intense enhanced absorption and emission bands detectable in single-scan ^{13}C measurement, and is obtained with proton decoupling in natural isotopic abundance. (In the absence of CIDNP enhancements accumulation of 200–600 scans is essential to obtain good spectra.)

The emission and enhanced absorption peaks of Fig. 3.6 have been assigned as follows:

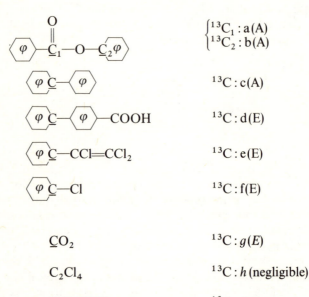

The observed polarizations are consistent with the accepted mechanism of the decomposition of benzoyl peroxide, and arise from initial sorting encounters between phenyl and benzoyloxy radicals within an initial solvent cage (with the heteronuclear decoupling, only net effects are observable but the CKO theory may be applied to this net polarization). LIPPMAA et al. [66] have reported a stronger emission for CO_2 during this thermal decomposition. (This is presumably due to the decreased solubility of CO_2 in tetrachloroethylene.) The emission from trichlorostyrene is a case of memory effect in which a polarized phenyl radical adds to C_2Cl_4 followed by fast loss of a chlorine atom to yield the polarized product.

The following characteristics of ^{13}C CIDNP should be stressed:

i) ^{13}C CIDNP signals of very dilute species can often be detected after a single scan at natural abundance. The ^{13}C polarization can be much stronger than the proton one. For example, during the decomposition of dibenzoyl peroxide LIPPMAA et al. [102j] report an enhanced absorption line corresponding to the $^{13}C_2$ of phenyl benzoate (see earlier formulae) and 2×10^4 times larger than the

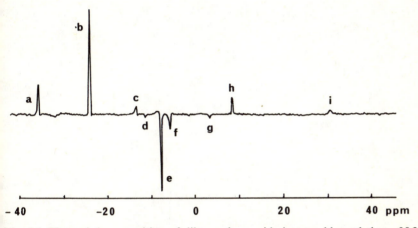

Fig. 3.6. Thermal decomposition of dibenzoyl peroxide in tetrachloroethylene: 25.2 MHz, ^{13}C spectrum, 15 accumulated scans, reference: $^{13}C\,C_6H_6$ (see text for the assignment of the peaks) (From SCHULMAN et al. [127])

corresponding signal observed at thermal equilibrium. These large polarizations probably result from the proximity of nuclei and electrons. In the example cited, the unpaired electron in the radical precursor is located directly above the ^{13}C atom detected.

ii) Polarized species can be detected by ^{13}C CIDNP when ^1H CIDNP is not available, for example when there is no proton in the molecule (CO_2) or when the protons are far from the single electron in the radical [185].

iii) ^{13}C CIDNP has the same advantages as ^{13}C NMR. The assignments of the peaks are easier with ^{13}C NMR than with ^1H NMR because chemical shifts are greater, there is no coupling between ^{13}C and ^{13}C (its natural abundance is too low), and coupling with protons is removed by decoupling techniques [1a, 1e, 1f].

iv) As mentioned by SCHULMAN et al. [127] CIDNP on ^{13}C does not necessarily appear on the carbon bound to the protons which show the CIDNP effect, even when protons are not decoupled.

v) With decoupling only the net effect is detected, but without decoupling all CIDNP effects may be observed. LIPPMAA et al. [66], to our knowledge, are the only authors who have reported non-decoupled CIDNP spectra.

7. CIDNP of ^{19}F

The first example of CIDNP in the ^{19}F NMR spectrum was reported in 1970 [53], for coupling products from the reaction of p-fluorobenzyl chloride and n-butyllithium. An intense polarization was observed in the NMR spectra of the major products, p,p'-difluorobibenzyl (I) and p-fluoropentyl-benzene (II):

$$^{19}F\text{—}C_6H_4\text{—}CH_2\text{—}CH_2\text{—}C_6H_4\text{—}^{19}F, \tag{I}$$

$$^{19}F\text{—}C_6H_4\text{—}(CH_2)_4\text{—}CH_3. \tag{II}$$

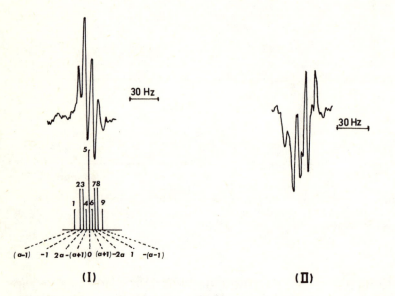

Fig. 3.7. Polarized ^{19}F spectra for the coupling products from the reaction of p-fluorobenzyl chloride and n-butyllithium.
(I) $^{19}F\text{—}C_6H_4\text{—}CH_2\text{—}CH_2\text{—}C_6H_4\text{—}^{19}F$ (II) $^{19}F\text{—}C_6H_4\text{—}(CH_2)_4\text{—}CH_3$
(From RAKSHYS [53])

These spectra show complex multiplet effects for both products (Fig. 3.7). For product (I) the phase of polarization changes at least five times. p-Fluorophenyl corresponds to an AA'BB'X spin system in which the ^{19}F nucleus is coupled to two nonequivalent nuclei in *ortho* and *meta* configurations. The ^{19}F NMR spectrum is a triplet of triplets.

RAKSHYS [53] has used the CKO theory with the superposition of the two following effects: E/A for the triplet ^{19}F $(H_{meta})_2$ and A/E for the triplet ^{19}F $(H_{ortho})_2$ to explain the polarized ^{19}F NMR spectra.

CIDNP of ^{19}F differs from CIDNP of proton in the two following characteristics:

i) the polarization can involve a nucleus far from the reaction site:

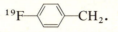

(consequence of spin delocalization in the benzyl radical);

ii) in the experiments reported by RAKSHYS, only the ^{19}F nucleus shows CIDNP effect and polarized ^{1}H spectra were not observed. Generally speaking, the polarization of the ^{1}H nucleus is not always observed when the ^{19}F nucleus show a CIDNP effect.

The greater sensitivity of ^{19}F to the CIDNP effect as compared to ^{1}H is probably the result of its participation in the π-orbital containing the unpaired electron.

In 1971 BARGON [88, 102e] reported ^{19}F CIDNP during homolytic aromatic substitution reactions. RAKSHYS [85] observed ^{19}F CIDNP during the reaction of p-fluorobenzyl chloride with sodium naphtalene (but no polarized ^{1}H nucleus was observed), and demonstrated that both benzyl chlorides and iodides react mainly by a carbanion route:

$$^{19}F\!-\!\!\left\langle\bigcirc\right\rangle\!\!-\!CH_2\cdot\ \cdot Naph^{\ominus}$$

The ^{19}F CIDNP of α-pentafluorobenzoyloxycyclohexane, the product of the thermal decomposition of pentafluorobenzoyl peroxide in cyclohexane has been studied by KOBRINA et al. [107]. BETHELL et al. [118] have observed ^{19}F CIDNP for the 1,2,2-triphenylethylfluoride product of the reaction of diphenylmethylene with benzylfluoride (polarized ^{1}H nucleus observed) or α-bromobenzyl fluoride (no polarized ^{1}H nucleus observed).

^{19}F CIDNP has been also reported for the products of insertion of diaryl-methylenes into benzylfluoride [155a] and during the dimerization of α-fluoro-benzyl radicals [155b].

8. CIDNP of ^{31}P and ^{15}N

The chemical polarization of ^{31}P was detected for the first time by LEVIN et al. [69] in the thermal decomposition of di-tert-butyl peroxide and dibenzoyl peroxide in the presence of alkyl phosphites. For instance, the pyrolysis of $(Me_3CO)_2$ in the presence of $(BuO)_3P$ gives rise to negative polarization of ^{31}P in the main products, $MeP(O)(OBu)_2$ and Bu_3PO_4.

Polarized ^{31}P NMR spectra have been reported during the free radicals reactions of alkyl and aryl phosphites [199–201] induced by thermolysis or photolysis of various compounds (peroxides, phenylazotriphenylmethane, acetone, etc.). In contrast to the ^{1}H CIDNP, the phosphorus nuclei give rise to polarization only via the transfer reactions of the parent radicals, but ^{31}P CIDNP can nevertheless, be easily observed.

The ^{31}P nucleus is very suitable for CIDNP studies, due to the very wide range of chemical shifts in its NMR spectra, its rather long nuclear relaxation times, and the extremely high values of hyperfine splittings on ^{31}P in many of the most important phosphorus-containing free radicals [199]. The theory of CIDNP effect via S–T transitions in the radical pairs is applicable to the ^{31}P CIDNP and explains all the principal features of polarized NMR spectra.

LIPPMAA et al. have observed a nuclear polarization of the ^{15}N nucleus in the thermal decomposition of diazoaminobenzene (1,3 triphenyltriazene) [102j, 186] and in diazo coupling reactions [102j, 188] (chemical polarization of both ^{15}N and ^{13}C have been reported for these reactions).

During the thermal decomposition of diazoaminobenzene, the most intense polarized lines in the ^{15}N spectrum result from the compounds that are also strongly polarized in the ^{13}C spectrum: aniline and diphenylamine. Polarization of the starting material has been also detected (only for N–2 of diazoaminobenzene: Ph—NH—^{15}N═N—Ph, no ^{13}C polarization appears in this compound). The general pattern of ^{15}N polarization agrees with that of ^{13}C and with the proposed reaction scheme. ^{15}N CIDNP is a welcome addition to both ^{1}H and ^{13}C CIDNP. Labelled samples of diazoaminobenzene with 44% of ^{15}N in either the central or the adjacent 1,3-positions have been used for these experiments, but ^{15}N CIDNP is not confined to enriched compounds only, and ^{15}N line of polarized aniline formed in the thermal decomposition of diazoaminobenzene with natural abundance of all nuclei has been observed.

II. Determination of the Magnetic Properties of Radicals and Molecules

The signs and amplitudes of nuclear polarization depend on the chemical parameters and magnetic properties of the free radicals involved in the reactions. If the chemical parameters are known, the radical properties can be determined from CIDNP patterns. CIDNP may be used as a tool to measure the magnetic properties of free radicals.

1. Sign Determination

Qualitatively, from Kaptein's rules [Eq. (39) of Chapter 2], the CIDNP effect depends on the sign of the following magnetic parameters:

$$\Delta g = g_1 - g_2: \text{ difference of } g \text{ factors, from } A \text{ or } E \text{ only,}$$

$$A_i: \text{ hyperfine coupling constant,}$$

$$J_{ij}: \text{ nuclear spin–spin coupling, from } A/E \text{ or } E/A \text{ only.}$$

The sign of one quantity can be deduced if the others are known. The only use for these data has been to determine the sign of the hyperfine coupling constant. This determination is often better than that by ESR.

FISCHER [76] has found a positive sign for the hyperfine coupling of H in CH_3CO. $(A_H = +5.1\,G)$. BRINKMAN et al. [198], ROTH et al. [159] and BARGON et al. [216] have determined the absolute signs of ^{19}F hyperfine coupling constants in fluorinated aromatic systems. This constant is positive for ^{19}F in ortho and para positions and negative for ^{19}F in meta position.

2. Hyperfine Coupling Constant A

Quantitative measurements have been obtained using the value of the enhancement factor V_{mn} [Eq. (33)]. This quantity is a function of several magnetic properties: A_i, Δg, J, T_1, which can be calculated if all the other parameters are known or correctly estimated. This is the method used by LEHNIG and FISCHER [48] to determine A_H in the .$CHCl_2$ radical. They have found $A_H = -17.0 \pm 1\,G$, but MORRIS et al. [117] using their simulation programs, have found a value between -13 and $-15\,G$. This result cannot be stated more precisely because the spectrum does not depend strongly on the A_H value, but they found that $-17\,G$ does not lead to the best fit. As the .$CHCl_2$ radical has not so far been detected by ESR such a measurement may be very useful to determine the magnetic properties of short-lived radicals in low concentration.

3. Spectroscopic Splitting Factor g or Δg

This factor is generally determined by ESR. Nevertheless, CIDNP is very sensitive to Δg values and may provide a means of determination of this parameter. As its measurement depends on the knowledge of other quantities, not always determined with precision, the values obtained are not very accurate.

The g factors of benzyl and benzhydryl systems have been studied by CLOSS and TRIFUNAC [43, 83b]. By means of different p-halogenated combinations they have examined the following pair:

$$ (p-X-C_6H_4)_2-CH\,.\,.CH_2-C_6H_4-p-Y $$

where X and Y are H, Cl, and Br. Assuming a g value for the benzyl radical of 2.0025, as given by ESR measurements, g factors have been obtained from CIDNP patterns [83b]. The graphical method for the semi-quantitative analysis of a first-order multiplet proposed by MULLER [109] outlined in 2.3.3.d has been used for this determination. The following values were obtained for the benzhydryl radical [83b] (p. 55):

X	CIDNP	ESR
H	2.0025	2.0025 (assumed)
Cl	2.0028	2.0030
Br	2.0049	2.0050

CLOSS et al. [42] and ADRIAN [73b] have studied the following system:

$$p\text{---}X\text{---}C_6H_4\text{---}\overset{\displaystyle |}{\underset{\displaystyle OH}{CH}}. .CH\text{---}(C_6H_4\text{---}p\text{---}Y)_2$$

where:
  a) $X = Br$, $Y = H$
  b) $X = Cl$, $Y = H$
  c) $X = Y = H$
  d) $X = H$, $Y = Cl$
  e) $X = H$, $Y = Br$.
  They found for Δg:

	a	b	c	d	e
[42]	2.7×10^{-3}	1.5×10^{-3}	0.47×10^{-3}	-0.33×10^{-3}	-2.7×10^{-3}
[73b]	1.77×10^{-3}	1.29×10^{-3}	0.48×10^{-3}	-0.32×10^{-3}	-1.77×10^{-3}

The differences arise from the choice of the model.

From calculated and observed CIDNP spectra taken during the photolysis of dibenzoyl peroxide (see 3.1.2) LEHNIG and FISCHER [48] have determined the g factor of $.CHCl_2$:

$$g(.CHCl_2) = 2.0080 \pm 0.0003 .$$

SHINDO et al. [96] have determined the ratio $\beta_e H_0 \Delta g/A$ for o-xylene and o-chlorotoluene assuming that $2|J|$ and $\beta_e g H_0 \gg A$. They found the values -0.168 and -0.053 respectively.

These determinations are not always precise. CLOSS [83b] reported the Δg value of 3.6×10^{-3} for

$$\boxed{\begin{array}{cc} C_6H_5-C. & .CH-C_6H_5 \\ \parallel & \vert \\ O & CH_3 \end{array}}$$

This value is not in agreement with the careful determination by ESR which gives $2.0 \pm 0.2 \times 10^{-3}$. The difference may be attributed to the neglect of relaxation. This makes it clear that all parameters of the system must be known and that the radical pair must be well described by the model used.

KAPTEIN et al. [81, 130] have used ^{13}C-enriched acetyl peroxide to measure g in the acetoxy radical. The value of: Δg is 0.0032 ± 0.0002 for the methyl/acetoxy radical pair. If $g_{methyl} = 2.0026$ then $g_{acetoxy} = 2.0058$.

4. J Determination (Overlap Integral)

This parameter is not often determined and its value is generally estimated. CLOSS et al. [42, 43] assumed a value of 10^8 radians/sec. J is determined to within an order of magnitude from the simulation of the spectrum. CLOSS [83b] shows, for instance, that for the benzyl and benzhydryl system previously mentioned, it is impossible to arrive at a set of consistent and reasonable g factors if J is greater than $0.2|A|$. KAPTEIN [81, 128a] has determined J in the case of the thermal decomposition of propionyl peroxide by means of simulation of the CIDNP spectra.

The best value was $J = 6 \times 10^8$ radians/sec, but higher values did not greatly perturb the spectrum. The limit is just $|J| > 10^8$ radians/sec. A similar result was observed by MORRIS et al. [117] who simulated the field dependence of the NMR spectrum of $Cl_2CHCHClCOOH$ formed in reactions of the nongeminate $Cl_2CH.$ and $.CHClCOOH$ radicals obtained by FISCHER and LEHNIG [75]. They found that $J = 0$ gives a better fit than $J = -2 \times 10^8$ radians/sec but these two values probably lie within the range of experimental error. Significant deviations arise when $J = -5 \times 10^8$ radians/sec.

With the same procedure, SHINDO et al. [96] have calculated J for the photoadducts of phenanthraquinone with o-xylene and o-chlorotoluene. They found respectively $J = +7.31 \times 10^9$ and $J = +6.61 \times 10^9$ radians/sec. These values have a sign and an order of magnitude which are different from previous results. This difference may be explained by a $\pi-\pi$ electron interaction (see 3.1.3).

5. Relaxation Times

We have seen previously (see 2.6.1) that there are two longitudinal relaxation times; T_1 is the NMR relaxation time of the products, and T_{1R} is the relaxation time in the free radicals of the radical pair. We always have: $T_1 \gg T_{1R}$. These factors are often used to explain the difference between the results predicted by the simple theory and experimental results.

a) Relaxation Time in the Radical (T_{1R}). In a high field, for $T_0 - S$ mixing a spin selection occurs in the radical pair. There is an enrichment of some nuclear spin states in the cage product and a depletion of the same states in the free radicals. The polarizations of products formed by direct combination or disproportionation in the radical pair, and that of products derived from transfer or coupling reactions of radicals escaping the pair are given by the kinetic formulation of the radical-pair mechanism [46, 76]:

$$
\begin{array}{c}
\underline{R_1} . . \underline{R_2}
\xrightarrow{k_c} \underline{R_1} - \underline{R_2} \qquad \text{(c)} \\[1em]
\xrightarrow{k_e} \underline{R_1} . + \underline{R_2} . \xrightarrow{1/T_{1R}} R_1 . + R_2 . \\[1em]
\downarrow \quad + SX \text{ or } R. \\[1em]
\underline{R_1} X, \underline{R_2} X, \underline{R_1} - R, \underline{R_2} - R \qquad \text{(e)}
\end{array}
$$

$$
\frac{(V_{mn})_e}{(V_{mn})_c} = - \frac{F}{F + 1/T_{1R}} \frac{k_c}{k_e}
$$

V_{mn} is the theoretical enhancement factor per product molecule formed, $F = k_1 (SX)$ (transfer reaction) or $F = k_2 (R.)$ (combination reaction), and T_{1R} is a single relaxation time in radicals (an approximation if there is a multi-level spin system). If the chemical rate constants and chemical concentrations are known, T_{1R} can be calculated.

Quantitative comparisons of CIDNP effects for cage and escape products have been reported. CLOSS and TRIFUNAC [44] found $T_{1R} \simeq 3.5 \times 10^{-4}$ sec for the benzyl radical (for the benzhydryl radical: $T_{1R} < 10^{-4}$ sec), and LEHNIG and FISCHER [48] found $T_{1R} \simeq 4.5 \times 10^{-4}$ sec for $.CHCl_2$.

WALLING and LEPLEY have proposed a "radical flux" model for the decomposition of phenylacetyl peroxide in CCl_3Br and CCl_3SO_2Cl. The polarization of benzyl bromide (or chloride) occurs via sorting accompanying diffusive encounters of benzyl and CCl_3. radicals, and depends on the concentration of CCl_3Br or CCl_3SO_2Cl and on the relaxation time in polarized benzyl radicals:

$$
C_6H_5 - CH_2 . + .CCl_3 \longrightarrow C_6H_5 - C\underline{H}_2 . + .CCl_3
$$

$$
C_6H_5 - C\underline{H}_2 . \longrightarrow C_6H_5 - CH_2 .
$$

$$
C_6H_5 - C\underline{H}_2 . + RX \longrightarrow C_6H_5 - CH_2 - X + R.
$$

Kinetic formulation of this model leads to the relaxation time $T_{1R} \simeq 0.76 \times 10^{-4}$ sec for the benzyl radical.

b) Relaxation Time in the Product (T_1). This time may be measured by classical pulse methods [1a, 1b]. CIDNP, however, may be used to determine T_1 in a product. The most extensive measurements of this parameter are those of MARUYAMA et al. [95] who have determined the relaxation time of the methine proton in 1,2 adducts formed during the photolysis of phenanthraquinone in the presence of various hydrogen donors at several temperatures. They determined two relaxation times: one during the irradiation of the sample, called T_1, and one when the light is off, called T_1^*. The measurement is made by observing the increase in intensity of the NMR spectra when the light is on and the corresponding decrease in intensity when the light is off. They found the following results:

Hydrogen donors	°C	T_1^* (sec)	T_1 (sec)
Fluorene	30	2.1	6.8
	50	2.6	7.0
	70	6.6	9.4
Xanthene	30	1.0	3.6
	50	0.9	4.3
	70	3.0	4.6
Diphenylmethane	50	2.3	3.2
9,10-dihydroanthracene	50	2.6	4.4

The change of T_1 with temperature is normal for such compounds. The lowest values for T_1^* may be explained by the presence of paramagnetic species during the irradiation whose concentration must decrease with an increase of the temperature.

MULLER and CLOSS [114] have shown the importance of T_1 in photochemical reactions by means of the same procedure as used by MARUYAMA et al. [95], but they did not give any value for T_1.

Chapter IV. The Chemically Induced Dynamic Electron Polarization (CIDEP Effect)

I. Introduction

It is well known that photolysis of molecules produces radicals which give an electron spin resonance spectrum (ESR). The theory of this effect may be found elsewhere [1a, 1d, 1i]. As for the CIDNP effect in NMR radical chemical reactions may give rise to a similar effect in ESR spectra. The first observation of this effect [170a, 170b] appeared before CIDNP, but was not developed, and ESR parameters were determined only for the identification of the different radicals. However, the anomalous behavior of the radicals, especially H atoms, was reported. The first work cited as CIDEP was that of PAUL and FISCHER [170c] who studied the reduction and oxidation of propionic and isobutyric acids.

Emission and enhanced absorption appear in a different manner from that for CIDNP because ESR spectra have different characteristics from NMR spectra. An example of CIDEP for an H atom is given in Fig. 4.1.

Although CIDNP seems to be well understood, CIDEP was more difficult to interpret and satisfactory theories have been published only at the beginning of 1973 [170s, 171]. The different attempts will be presented below.

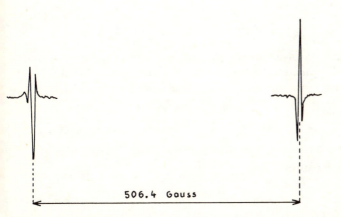

506.4 Gauss

Fig. 4.1. ESR lines of H atoms showing CIDEP effect. Note the inversion of the low-field lines. (From EIBEN and FESSENDEN [170j])

II. Experimental Procedures

As the ESR spectra are obtained in cells which are quite different from those used for NMR, and as irradiation is also different, the experimental procedures will be described.

1. Electron Irradiation

The electron beam irradiation is usually obtained from a VAN DE GRAFF accelerator [170a, 170b, 170h, 170i, 170j]. The beam must be directed along the axis of the magnetic field, because in other directions the electron trajectories are bent and the intensity of the beam in the cell could not be measured. All experimental spectrometers used for this purpose have a hole in one pole cap through which the electron beam is directed and enters the cavity through a thin wall or a hole [170j].

As ionization also occurs, the characteristics of the samples are rapidly modified. To minimize the effects of this, a flow system is used where the liquid is recycled by an external pump [170b, 170i]. In pulsed electron beam generators [170b, 170i] spectrometers are modified to study transient ESR spectra.

2. UV Irradiation

For UV irradiation, modifications are minor. The UV light is focused and passed through a small quartz window [170e, 170m]. ATKINS et al. [170f, 170u] used a nitrogen gas laser directed normal to H_0 to avoid light polarization. In his last paper, ATKINS [170u] gives some details on the observation of the transient ESR spectra.

3. Other Methods

The only non-irradiative measurement was made by PAUL and FISCHER [170c]. Here the oxidation and reduction occurred in the cell: reactants were mixed at the entrance of the cell and flowed through it. Variable concentration was used with constant flow rate. This technic is also applied by SHUVALOV et al. [170v].

III. Experimental Results

FESSENDEN and SCHULER [170a] measured the spectra of 25 organic radicals obtained from hydrocarbons by electron beam irradiation. These radicals were studied only from the ESR point of view and their spectroscopic constants are given. Anomalous intensities were observed only in the case of the ethyl radical and H and D atoms.

Later SMALLER et al. [170b], using a pulsed electron beam, measured the decay of the ESR signal and obtained the rate constants for cyclopentyl and cyclohexyl radicals as a function of temperature. They studied the effect of radical scavengers such as sulfur-containing compounds or oxygen on the radical recombination. They also found that the cyclopentyl radical showed anomalous intensities. Lines at low field had a reduced intensity while lines at high field had enhanced intensities, equivalent to an E/A effect. They suggested that this was due to a population inversion. They gave the experimental variation of the intensity of one line with time. It appeared clearly as an emission line followed by an absorption line in the steady state. The time when emission occurred was of the order of T_1. Such an effect was also found with other radicals.

PAUL and FISCHER [170c] mentioned the CIDEP effect in their study of the oxidation and reduction of propionic and isobutyric acids and were the first to give the enhancement factor for CIDEP.

Since then, several studies have been reported, especially for H and D atoms. NETA et al. [170h] studied the reaction of H and D (produced by an electron-beam generator in a solution of 0.1M perchloric acid) with 67 organic compounds. In all cases, the ESR spectrum of H$^{\cdot}$ showed a low-field emission line and a high-field enhanced absorption line. No interpretation was given, except that this effect is the result of an abnormal population of electronic states. SMALLER et al. [170i] reported similar experiments on 12 different compounds, but they found that the dissymmetry of the H spectrum disappears when the lifetime of the radical is less than 100 μsec. They also used Mn^{2+} ions to shorten the relaxation time and found that this abolished the CIDEP effect.

The CIDEP effect was also found with organic radicals. Photolysis of tartric acid [170e] gave several radicals, one of which showed only emission lines, while the others displayed both absorption and emission. In the latter case, when the pH of the solution was increased, emission disappeared and absorption was enhanced. On the other hand, at low pH values only emission lines were observed. Furthermore, emission was enhanced for the low-field part of the spectra. The only explanation for this is that the electron-spin population is not inverted, but there is a relatively large overpopulation in the upper state.

EIBEN and FESSENDEN [170j] in their study of aqueous solutions of organic compounds under electron irradiation, noticed abnormal intensity behavior, especially in the case of hydroxycyclohexadienyl radicals and benzoate ions, where the lowest-field lines were inverted. They mentioned that T_1 of these radicals is long, which may explain this effect.

ATKINS et al. [170f, 170q] with transient ESR measurements of solutions irradiated by laser pulses, observed emission in benzophenone, benzaldehyde, 2- and 4-chlorobenzaldehyde, acetophenone, dibenzyl and anthraquinone, but no effect on aliphatic molecules containing carbonyl groups. The theory will be developed [170f] in 4.4.

WONG and WAN report the ESR emission from the photochemically produced 1-4-naphthosemiquinone radical [170m] from the durosemiquinone-phenoxy radical pair [170p] and from benzophenone [170y]. They propose an interpretation [170n, 170r] which will be discussed later.

GLARUM and MARSHALL [170d] irradiated solutions of K, Rb and Cs in dimethoxyethane. They observed the ESR spectrum from the solvated e^-. Rubidium solutions gave rise to emission lines. It was demonstrated experimentally that this effect was not due to light polarization. This theory will be outlined later.

In these studies, a CIDEP effect was observed but generally not explained theoretically. SHUVALOV et al. [170v] have studied the system Ti(III)-H_2O_2-isobutyric acid or isopropanol. Their results seem to be in contradiction with Adrian's model [170g] and 4.4.1. CHEN et al. [203] have studied the photochemistry of propionaldehyde using ESR and NMR. From their experimental results, interpreted with the theory of TOMKIEWICZ et al. [122], they suggest different photochemical reaction steps.

FISCHER [20] has also studied ESR spectra of transient alkyl radicals to measure hyperfine coupling constants in order to interpret CIDNP spectra. No CIDEP effect, however, was reported.

IV. The Theory of the CIDEP Effect

Several authors have outlined methods of approach to a general theory mostly starting with the CIDNP theory.

1. Kaptein-Adrian Theory

The first attempt was made by KAPTEIN and OOSTERHOFF [23] who started from the radical-pair model and applied it to the case of H˙. This idea was later developed by ADRIAN [104, 170g, 170o, 170v, 170x], who applied it to all radicals. At the same time, ATKINS et al. [170k] showed that spin–rotation coupling, as defined in 2.1.4b Eq. (8), plays an important role.

KAPTEIN and ADRIAN started with the equations previously given in 2.2.2 which describe the behavior of the spin system in the radical pair. To obtain the spin state of the electron of a radical after an escape, we may calculate the unpaired electron spin density ϱ of this radical.

ϱ is given by:

$$\varrho(t) = \langle \psi^*(t) | S_{z_1} - S_{z_2} | \psi(t) \rangle$$

where ψ is the same function as in 2.2.2 Eq. (22). S_{z1} and S_{z2} are components of S_1 and S_2 in the direction of the external magnetic field, which lies along the z axis, for electrons 1 and 2.

If $\varrho > 0$, the electron spin is parallel to the external magnetic field and hence in the upper Zeeman level. We obtain an emission line in the ESR spectra. If $\varrho < 0$, we have an absorption line.

Using the expression of ψ we obtain:

$$\varrho(t) = c_{T_0}(t)\, c_S^*(t) + c_{T_0}^*(t)\, c_S(t).$$

We may then write the expression of ϱ using the value of $c_{T_0}(t)$ and $c_S(t)$ given by (23) and (24) of Chapter 2. The value of ϱ observed is the mean value $\bar{\varrho}$ of $\varrho(t)$ over a long time. The result depends on the models used to describe radical-pair behavior.

KAPTEIN[23] used the cage model (see 2.3.1.a) and found a selectivity parameter $\Delta_\pm$ given by:

$$\Delta_\pm = \pm\, \frac{AJ\tau_c^2}{1 + 4\omega^2\tau_c^2}$$

where the variables have their usual meaning, $+$ refers to an α nuclear spin state and $-$ to a β nuclear spin state. The population of the different spin states $\alpha_e\alpha_N$, $\alpha_e\beta_N$, $\beta_e\alpha_N$, $\beta_e\beta_N$ of H$\cdot$ may be expressed as $k/4(1+\Delta_+)$, $k/4(1+\Delta_-)$, $k/4(1-\Delta_+)$, and $k/4(1-\Delta_-)$ where k is a chemical rate constant. With this theory KAPTEIN [23] explained the result observed with H$\cdot$ atoms.

ADRIAN [170g] used the diffusion model of NOYES with Eq. (12) and assumed that $J = 0$ during the time of separation between encounters and $|J| \gg \mathcal{H}_m$ during an encounter. This led to:

$$\varrho(|J|, \tau_c') = 0.85(\mathcal{H}_m J/|\mathcal{H}_m J| \times \sqrt{\mathcal{H}_m \tau_D}\,[|c_S(0)|^2 - |c_{T_0}(0)|^2]\, \sin 2|J|\tau_c')$$

where τ_D is the mean time between diffusive displacements of the radical and τ_c' the duration of a typical collision between the radicals, for instance, the time when $J > \mathcal{H}_m$.

In this case only T_0 and S electron-spin states are used and not $T_\pm$. If it is assumed that

$$\langle \sin 2|J|\, \tau_c' \rangle_{(\text{collision average})} \simeq 0.1$$

with $\tau_D = 10^{-11}$ sec and $\mathcal{H}_m = 10^8$ radians/sec, it is found for the pure singlet or triplet state, that $|\langle\bar{\varrho}\rangle| = 2.7 \times 10^{-3}$. This is greater than the difference of populations occurring in ESR at 9000 MHz, which is of the order of 1.4×10^{-3}.

Nevertheless, radicals can be produced from the collision of independent radicals with uncorrelated electron spins. In this case, it is assumed that the probability of radical combination is

$$P_R = k_r |c_S(0)|^2$$

where k_r is the rate constant for product formation. If $J < 0$ and $T_\pm$ are neglected, we obtain:

$$\langle\!\langle \varrho \rangle\!\rangle_{uncor} = 0.021 \, k_r \left(\frac{\mathscr{H}_m}{|\mathscr{H}_m|} \right) \sqrt{\mathscr{H}_m \tau_D} \, .$$

Later, ADRIAN [104] included $T_\pm$ spin states. But in this case, as shown by ATKINS et al. [170k], spin rotation may play an important role and may be introduced in the matrix of the magnetic interactions connecting singlet and triplet states. This term makes hyperfine splitting independent of polarization in the totally emissive ESR spectrum. But at present, no experiment can prove this assumption.

ADRIAN [170o] has refined this solution by using an exchange interaction in which $J(R) = J_0 \exp(-\lambda R)$, and calculating the average singlet–triplet splitting and the corresponding polarization with respect to various diffusive paths of the radicals. The results of this calculation are in better agreement with the experiments and confirm his original theory.

This theory explains the results of FISCHER [170c] in detail, but $T_\pm$ electron spin states must be taken into account in the totally emissive spectrum.

ATKINS et al. [170k], calculating the probability that a state that would contribute to a line is a singlet, have shown that the hyperfine coupling interactions give distorted emission spectra, but that spin rotation interactions do not perturb the binomial intensity distribution. His theory is developed in Ref. 176b. In this text the different terms which contribute to the CIDEP effect are analysed. They are: i) Spin-orbit processes which are very small and may be neglected. ii) Spin-rotation process which is important, and for which the intensity of the CIDEP effect is given. iii) Hyperfine process which comes from the existence of the hyperfine coupling constant A. The variation of the intensity of the ESR spectrum is calculated for this process. In his last paper [170x], ATKINS calculates the CIDEP effect using a $T_0 - S$ mixing which varies linearly or exponentially with time. $S - T_-$ mixing induced by an anisotropic g-factor may occur [170z].

Adrian's theory, which is a better approach to the CIDEP effect, uses a discontinuous model. PEDERSEN and FREED [170l] have used the stochastic Liouville method to calculated the spin-density matrix, including spin-dependent parameters and the effects of translational diffusion. As this theory cannot be developed analytically they have computed solutions and mentioned a number of trends, especially the influence of the diffusive coefficient D which is not directly used in previous theories.

In their latest publication PEDERSEN and FREED [170s] have used two expressions for $J(R) : J(R) = J_0 \exp[-\lambda(r-d)], r \gg d$ or $J(R) = J_0 \delta_{i0}$, where δ_{i0} is the Dirac function, which is different from zero only when the distance of the ith electron to the nucleus is equal to zero. This theory shows that the first expression of J is more satisfactory than the second one. As their calculations agree with the experiments, they have concluded that their theory, based on the Liouville equation given in terms of the spin-density matrix $\varrho(r_i, t)$ is the correct interpretation of CIDEP. Recently they have shown the influence of the variation of J on the CIDEP effect [170w]. The same idea was developed analytically by EVANS et al. [171] and has been presented in 2.5.3.

2. Glarum-Marshall Model

To interpret the important CIDEP effect observed from the Rb^- ions obtained by flash photolysis, GLARUM and MARSHALL [170d] assumed that the Zeeman electronic levels are differently populated when the light is absorbed and that the relaxation follows the dissociation. If $N^{\pm}$ represents the number of electrons in the upper and lower levels, respectively, and $I(t)$ the light intensity, then

$$dN^{\pm}/dt = \tfrac{1}{2}(1 \pm \gamma) I(t) \mp (k_1 N^+ - k_2 N^-)$$

where k_1 and k_2 determine the rate of thermal equilibration between electronic spin levels and are functions of T_1, and γ is a factor which determines the inequality of populations of α or β electron-spin states in the photochemical production of $Rb^{\cdot}$. If $\gamma = 0.5$, there is no CIDEP effect. The solution of this equation gives the

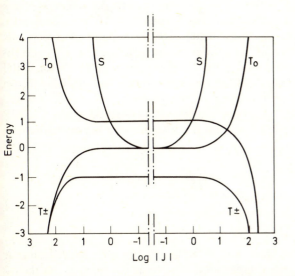

Fig. 4.2. Energy levels showing the dissociation into the laboratory field. The left- and right-hand plots are for $H_0 \perp z$ and $H_0 // z$, respectively. (From GLARUM and MARSHALL [170d])

intensity of the ESR signal as a function of time and can explain the experimental results. The ESR emission signal is thus dependent on light intensity.

This is verified experimentally. Measurements show $\gamma \geq 0.30$, which corresponds to 65 percent of electrons formed in the upper level. The problem of the origin of such a polarization of electrons by photolysis is not solved. A detailed investigation of the electronic energy levels of Rb^-, represented in Fig. 4.2 (cf. Fig. 2.4), leads to two alternative interpretations. (i) As e^- leaves Rb^- in the initial states, the value of J decreases and $S - T_0$ mixing occurs, due to spin-orbit coupling. Then pseudodipolar interactions mix T_0 and T_+. The two peripherical electrons of Rb^- are then in the α state and e^- which leaves Rb^- is also in the α state. (ii) If there is a difference in g values for the electrons in the dissociation products $Rb^{\cdot}$ and e^-, mixing occurs between singlet and triplet when $E = 0$. If the g value of e^- exceeds that of $Rb^{\cdot}$, adiabatic dissociation from the S state leads to an electron in an α state and the remaining electron on $Rb^{\cdot}$ is in a β state. But experiments do not indicate which of these two interpretations is correct. The Glarum-Marshall theory may explain some but not all features of CIDEP.

3. Simple Theory for the H Atom

Recently, HUTCHINSON et al. [170n] have proposed a theory for a one-electron atom. Their work is based on the following reactions

$$e_s^- + H^+ \rightleftarrows (H^+ e_s^-) \rightarrow H^{\cdot}$$

where the rate of reactions is determined by the second step. It is supposed that the complex $(H^+ e_s^-)$ is formed in a higher electronic state which cascades to the ground state $^2S_{1/2}$. From this scheme it is apparent that triplet sublevels of the ground state have more precursors than singlet sublevels. The probability for each step of the cascade is calculated and the sum gives the population of each ESR level. It is then easy to deduce the intensities of the ESR spectra. Their results represent a qualitative interpretation of the CIDEP spectra of $H^{\cdot}$ [170h, 170j] and $D^{\cdot}$ [170i] atoms. This theory has been used to explain the CIDEP effect in radicals containing carbonyl groups [170r].

As we can see, the CIDEP theory is not as well understood as the CIDNP theory. There are not enough experimental results and more measurements are needed to test the theories.

Conclusion

CIDNP was discovered in 1967 and interpreted in 1969–1970 by the radical-pair theory. It is now reasonably well understood. The basic concepts of the theory are well established, and the different treatments proposed by several authors differ only slightly in quantitative formulation.

At present, efforts are being made to develop a general theory covering the amplitudes and signs of polarizations observed in high- and low-field experiments. This will necessitate new experiments for some particular cases which have not been completely investigated, for instance, reactions in low magnetic field, biradical reactions etc.

As shown in Chapter 3 CIDNP is a powerful tool for the determination of reaction mechanisms and radical properties. The list of reactions studied, mechanisms elucidated and radical parameters determined by CIDNP will certainly increase rapidly. Some of the most promising applications are for fast-reaction kinetics, structural problems of short-lived radicals and determination of magnetic parameters (such as nuclear relaxation times).

Quantitative studies of CIDNP utilizing pulsed NMR techniques (Fourier transform NMR), which are very suitable for these studies, and CIDNP of nuclei other than the proton (especially ^{13}C) will undoubtedly be very interesting.

Quantitative aspects of CIDNP depend on the theory of diffusion in liquids, and this effect may be a valuable tool for the study of the microscopic behavior of radicals in solution. Theoretical aspects of CIDNP in relation to the behavior of two neighboring radicals remain to be elucidated, for instance: How far apart must the radical components be from each other for the singlet–triplet mixing? How much polarization will there be in any given reaction? The latter question is important in order to know how much of a given reaction is actually occuring through a radical mechanism involving an intermediate with a polarizing radical pair. In fact, minor reaction pathways often give rise to a larger polarization than the main reaction.

The CIDEP is not so developed as CIDNP. Its elucidation requires more experiments, and these will certainly also throw light on the CIDNP effect.

Acknowledgements. It is a great pleasure for us to thank Professors J. BARRIOL, M. NICLAUSE, R. MARTIN and M. BACK for their encouragement and their valuable remarks. We greatly appreciate the collaboration of Mrs. N. GROOS in drawing the figures.

One of us (C.R.) is indebted to the Centre National de la Recherche Scientifique for its financial support. We are grateful to the journal publishers and the respective authors for permission to reproduce their figures in this book.

Addendum. As this book was in press, we have met Professor ERNST who have mentionned an important point about the use of the Fourier transformed spectroscopy in CIDNP experiments 217. As CIDNP is concerned with non-equilibrium states of nuclear spin systems, the relative signal intensities obtained from Fourier experiments depend on the flip angle used for the pulses in the case of homonuclear strong-coupled systems, (second order spectra). This is the case for ^1H, ^{19}F. This situation may also be met for fully enriched molecules with ^{13}C, ^{15}N. The phenomenon is presented on an *AB* spectrum with pur multiplet effect 217. This difficulty does not affect the observation of a net effect since all lines are equally perturbed, but is of importance with a multiplet effect. For low natural abundant nuclei such as ^{13}C, which is observed with decoupling, CIDNP effect is not perturbed by this phenomenon. This can be explained by the fact that only net effect is observed in this case (see p. 97). To observe CIDNP with multiplet effect using Fourier transformed spectroscopy small flip angles are necessary (about 10–20°).

References

1. a) CARRINGTON, A., MC LACHLAN, A. D.: Introduction to Magnetic Resonance. New-York: Harper and Row 1967
   b) ABRAGAM, A.: The Principles of Nuclear Magnetism. London: Oxford University Press 1962; or ABRAGAM, A.: Les Principes du Magnétisme Nucléaire. Paris: Presses Universitaires de France 1961.
   c) CORIO, P. L.: Structure of High Resolution NMR spectra. New York: Academic Press 1966
   d) SLICHTER, C. P.: Principles of Magnetic Resonance. New York: Harper and Row 1963.
   e) EMSLEY, J. W., FEENEY, J., SUTCLIFFE, L. H.: High Resolution Nuclear Magnetic Resonance Spectroscopy, Vol. I and II. Oxford: Pergamon Press 1965.
   f) HOFFMANN, R. A., FORSEN, S., GESTBLOM, B.: NMR. Basic Principles and Progress, Grundlagen und Fortschritte, Vol. 5 (Eds.): DIEHL, P., FLUCK, E., KOSTFELD, R.: Berlin-Heidelberg-New York: Springer 1971.
   g) NOGGLE, J. H., SCHIRMER, R. E.: The Nuclear Overhauser Effect. New York: Academic Press 1971.
   h) MOREAU, G.: Bull. Soc. Chim. France **1969**, 1770.
   i) PAKE, G. E.: Paramagnetic Resonance. New York: W. A. Benjamin 1962.
   f) VAN VLECK, J. H.: The Theory of Electric and Magnetic Susceptibilities. London: Oxford University Press 1948.
2. a) BARRIOL, J.: Elements of Quantum Mechanics with Chemical Applications. New York: Barnes and Noble Inc. 1971; or BARRIOL, J.: Eléments de Mécanique Quantique. Paris: Masson et Cie 1966.
   b) MESSIAH, A.: Quantum Mechanics, Vol. I. Amsterdam: North Holland Publ. Co. 1967; or MESSIAH, A., Mécanique Quantique, Vol. I. Paris: Dunod 1964.
   c) SCHIFF, L. I.: Quantum Mechanics. New York: McGraw-Hill 1955.
3. ITOH, K., HAYASHI, H., NAGAKURA, S.: Mol. Phys. **17**, 561 (1969).
4. NOYES, R. M.: J. Chem. Phys. **22**, 1349 (1954).
5. BARGON, J., FISCHER, H., JOHNSEN, U.: Z. Naturforsch. **22a**, 1551 (1967).
6. BARGON, J., FISCHER, H.: Z. Naturforsch. **22a**, 1556 (1967).
7. WARD, H. R.: J. Am. Chem. Soc. **89**, 5517 (1967).
8. WARD, H. R., LAWLER, R. G.: J. Am. Chem. Soc. **89**, 5518 (1967).
9. LAWLER, R. G.: J. Am. Chem. Soc. **89**, 5519 (1967).
10. BARGON, J., FISCHER, H.: Z. Naturforsch. **23a**, 2109 (1968).
11. KAPTEIN, R.: Chem. Phys. Letters **2**, 261 (1968).
12. LEPLEY, A. R.: J. Am. Chem. Soc. **90**, 2710 (1968).
13. COCIVERA, M.: J. Am. Chem. Soc. **90**, 3261 (1968).
14. WARD, H. R., LAWLER, R. G., LOKEN, H. Y.: J. Am. Chem. Soc. **90**, 7359 (1968).
15. Chem. Eng. News **46**, 40 (Jan. 15, 1968).
16. CLOSS, G. L., CLOSS, L. E.: J. Am. Chem. Soc. **91**, 4549 (1969).
17. CLOSS, G. L., CLOSS, L. E.: J. Am. Chem. Soc. **91**, 4550 (1969).
18. CLOSS, G. L.: J. Am. Chem. Soc. **91**, 4552 (1969).
19. CLOSS, G. L., TRIFUNAC, A. D.: J. Am. Chem. Soc. **91**, 4554 (1969).
20. FISCHER, H.: J. Phys. Chem. **73**, 3834 (1969).
21. FISCHER, H., BARGON, J.: Acc. Chem. Res. **2**, 110 (1969).
22. LEHNIG, M., FISCHER, H.: Z. Naturforsch. **24a**, 1771 (1969).
23. KAPTEIN, R., OOSTERHOFF, J. L.: Chem. Phys. Letters **4**, 195 and 214 (1969).

24. WARD, H. R., LAWLER, R. G., COOPER, R. A.: Tetrahedron Letters 527 (1969).
25. WARD, H. R., LAWLER, R. G., COOPER, R. A.: J. Am. Chem. Soc. **91**, 746 (1969).
26. WARD, H. R., LAWLER, R. G., LOKEN, H. Y., COOPER, R. A.: J. Am. Chem. Soc. **91**, 4928 (1969).
27. LEPLEY, A. R.: Chem. Commun. 64 (1969).
28. LEPLEY, A. R.: Chem. Commun. 1460 (1969).
29. LEPLEY, A. R., LANDAU, R. L.: J. Am. Chem. Soc. **91**, 748 (1969).
30. LEPLEY, A. R.: J. Am. Chem. Soc. **91**, 749 (1969).
31. LEPLEY, A. R.: J. Am. Chem. Soc. **91**, 1237 (1969).
32. RUSSEL, G. A., LAMSON, D. W.: J. Am. Chem. Soc. **91**, 3967 (1969).
33. SCHÖLLKOPF, U., OSTERMANN, G., SCHOSSIG, J.: Tetrahedron Letters **1969**, 2619.
34. SCHÖLLKOPF, U., LUDWIG, U., OSTERMANN, G., PATSCH, M.: Tetrahedron Letters **1969**, 3415.
35. LANE, A. G., RUCHARDT, C., WERNER, R.: Tetrahedron Letters **1969**, 3213.
36. JEMISON, R. W., MORRIS, D. G.: Chem. Commun. **1969**, 1226.
37. MORRIS, D. G.: Chem. Commun. **1969**, 1345.
38. BALDWIN, J. E., BROWN, J. E.: J. Am. Chem. Soc. **91**, 3647 (1969).
39. RIEKER, A., NIEDERER, P., LEIBFRITZ, D.: Tetrahedron Letters **1969**, 4287.
40. RYKOV, S. V., BUCHACHENKO, A. L., BALDIN, V. I.: J. Struct. Chem. **10**, 814 (1969).
41. CLOSS, G. L., TRIFUNAC, A. D.: J. Am. Chem. Soc. **92**, 2183 (1970).
42. CLOSS, G. L., DOUBLEDAY, C. E., PAULSON, D. R.: J. Am. Chem. Soc. **92**, 2185 (1970).
43. CLOSS, G. L., TRIFUNAC, A. D.: J. Am. Chem. Soc. **92**, 2186 (1970).
44. CLOSS, G. L., TRIFUNAC, A. D.: J. Am. Chem. Soc. **92**, 7227 (1970).
45. CLOSS, G. L., PAULSON, D. R.: J. Am. Chem. Soc. **92**, 7229 (1970).
46. FISCHER, H.: Z. Naturforsch. **25**a, 1957 (1970).
47. FISCHER, H.: Chem. Phys. Letters **4**, 611 (1970).
48. LEHNIG, M., FISCHER, H.: Z. Naturforsch. **25**a, 1963 (1970).
49. Chem. Eng. News **48**, 36 (March 9, 1970).
50. KAPTEIN, R., DEN HOLLANDER, J. A., ANTHEUNIS, D., OOSTERHOFF, L. J.: Chem. Commun. **1970**, 1687.
51. IWAMURA, H., IWAMURA, M., NISHIDA, T., MIURA, I.: Bull. Chem. Soc. Japan **43**, 1914 (1970).
52. IWAMURA, H., IWAMURA, M., TAMURA, M., SHIOMI, K.: Bull. Chem. Soc. Japan **43**, 3638 (1970).
    IWAMURA, H., IWAMURA, M.: Tetrahedron Letters **1970**, 3723.
53. RAKSHYS, Jr., J. W.: Chem. Commun. **1970**, 578.
54. JEMISON, R. W., MAGESWARAN, S., OLLIS, W. D., POTTER, S. E., PRETTY, A. J., SUTHERLAND, I. O., THEBTARANONTH, Y.: Chem. Commun. **1970**, 1201.
55. a) GARST, J. F., COX, R. H., BARBAS, J. T., ROBERTS, R. D., MORRIS, J. I., MORRISON, R. C.: J. Am. Chem. Soc. **92**, 5761 (1970).
    b) GARST, J. F., COX, R. H.: J. Am. Chem. Soc. **92**, 6389 (1970).
56. WARD, G. A., CHIEN, J. C. W.: Chem. Phys. Letters **6**, 245 (1970).
57. LEPLEY, A. R., COOK, P. M., WILLARD, G. F.: J. Am. Chem. Soc. **92**, 1101 (1970).
58. COCIVERA, M., TROZZOLO, A. M.: J. Am. Chem. Soc. **92**, 1772 (1970).
59. COCIVERA, M., ROTH, H. D.: J. Am. Chem. Soc. **92**, 2573 (1970).
60. BALDWIN, J. E., ERICKSON, W. F., HACKLER, R. E., SCOTT, R. M.: Chem. Commun. **1970**, 576.
61. HOLLAENDER, J., NEUMANN, W. P.: Angew. Chem. **82**, 813 (1970); – Angew. Chem. Int. Ed. (Engl.) **9**, 804 (1970).
62. SCHÖLLKOPF, U., HOPPE, I.: Tetrahedron Letters **1970**, 4527.
63. SCHÖLLKOPF, U.: Angew. Chem. **82**, 795 (1970); – Angew. Chem. Int. Ed. (Engl.) **9**, 763 (1970).
64. SCHÖLLKOPF, U., SCHOSSIG, J., OSTERMANN, G.: Liebigs Ann. Chem. **737**, 158 (1970).
65. OSTERMANN, G., SCHÖLLKOPF, U.: Liebigs Ann. Chem. **737**, 170 (1970).
66. LIPPMAA, E., PEHK, T., BUCHACHENKO, A. L., RYKOV, S. V.: Chem. Phys. Letters **5**, 521 (1970).
67. RYKOV, S. V., BUCHACHENKO, A. L., KESSENICH, A. V.: Spectroscopy Letters **3**, 55 (1970).

68. Buchachenko, A. L., Kessenikh, A. V., Rykov, S. V.: Zh. Eksperim. Teor. Fiz. **58**, 766 (1970); – Soviet Physics JETP **31**, 410 (1970).
69. Levin, Ya. A., Il'yasov, A. V., Pobedimskii, D. G., Gol'dfarb, E. I., Saidashev, I. I., Samitov, Yu. Yu.: Izv. Akad. Nauk. SSSR, Ser. Khim. **1970**, 1680.
70. Koenig, T., Mabey, W. R.: J. Am. Chem. Soc. **92**, 3804 (1970).
71. Jacobus, J.: Chem. Commun. **1970**, 709.
72. Ward, H. R., Lawler, R. G., Marzilli, T. A.: Tetrahedron Letters 521 (1970).
73. a) Adrian, F. J.: J. Chem. Phys. **53**, 3374 (1970).
    b) Adrian, F. J.: J. Chem. Phys. **54**, 3912 (1971).
74. Blank, B., Fischer, H.: Helv. Chim. Acta **54**, 905 (1971).
75. Fischer, H., Lehnig, M.: J. Phys. Chem. **75**, 3410 (1971).
76. Fischer, H.: Topics Curr. Chem. Fortschr. Chem. Forsch. **24**, 1 (1971).
77. Den Hollander, J. A., Kaptein, R., Brand, P. A. T. M.: Chem. Phys. Letters **10**, 430 (1971).
78. Kaptein, R., Frater-Schroder, M., Oosterhoff, L. J.: Chem. Phys. Letters **12**, 16 (1971).
79. Kaptein, R.: Chem. Commun. **1971**, 732.
80. Kaptein, R., Verheus, F. W., Oosterhoff, L. J.: Chem. Commun. **1971**, 877.
81. Kaptein, R.: Ph. D. thesis, Leiden (1971).
82. Closs, G. L.: J. Am. Chem. Soc. **93**, 1546 (1971).
83. 23rd International Congress of Pure and Applied Chemistry, Boston, USA, 26–30 July 1971, Vol. 4. London: Butterworths 1971.
    a) Blank, B., Mennitt, P. G., Fischer, H., p. 1.
    b) Closs, G. L., p. 19.
84. Rieker, A., Niederer, P., Stegmann, H. B.: Tetrahedron Letters **1971**, 3873.
85. Rakshys, Jr., J. W.: Tetrahedron Letters **1971**, 4745.
86. Kawazoe, Y., Araki, M.: Chem. Pharm. Bull. **19**, 1278 (1971).
87. Lawler, R. G., Ward, H. R., Allen, R. B., Ellenbogen, P. E.: J. Am. Chem. Soc. **93**, 789 (1971).
88. Bargon, J.: J. Am. Chem. Soc. **93**, 4630 (1971).
89. Charlton, J. L., Bargon, J.: Chem. Phys. Letters **8**, 442 (1971).
90. Baekelmans, P., Martens, G. J.: Int. J. chem. Kinet. **3**, 375 (1971).
91. Walling, C., Lepley, A. R.: J. Am. Chem. Soc. **93**, 546 (1971).
92. Fahrenholtz, S. R., Trozzolo, A. M.: J. Am. Chem. Soc. **93**, 251 (1971).
93. Papers presented at 10th International Symposium on Free Radicals, August 31, September 1–3, 1971, Lyons (France).
    a) Levit, A. F., Gragerov, I. P., Buchachenko, A. L.
    b) Fischer, H., Lehnig, M., Blank, B.
    c) Adrian, F. J.
94. Iwamura, H., Iwamura, M., Nishida, T., Yoshida, M., Nakayama, J.: Tetrahedron Letters **1971**, 63.
95. Maruyama, K., Shindo, H., Otsuki, T., Maruyama, T.: Bull. Chem. Soc. Japan **44**, 2756 (1971).
96. Shindo, H., Maruyama, K., Otsuki, T., Maruyama, T.: Bull. Chem. Soc. Japan **44**, 2789 (1971).
97. Maruyama, K., Otsuki, T.: Bull. Chem. Soc. Japan **44**, 2885 (1971).
98. Morris, D. G.: Chem. Commun. **1971**, 221.
99. Garst, J. F., Barton, F. E., Morris, J. I.: J. Am. Chem. Soc. **93**, 4310 (1971).
100. Baldwin, J. E., Brown, J. E., Hofle, G.: J. Am. Chem. Soc. **93**, 788 (1971).
101. Berson, J. A., Bushby, R. J., McBride, J. M., Tremelling, M.: J. Am. Chem. Soc. **93**, 1544 (1971).
102. Papers presented at International Colloquium on "Chemically Induced Dynamic Nuclear Polarization and Its Impact on the Study of Reaction Mechanisms", March 18–19, 1971, Brussels (Belgium). Ind. Chim. Belg. **36**, 1051 (1971).
    a) Fischer, H., p. 1054.
    b) Schöllkopf, U., p. 1057.
    c) Lemaire, H., Subra, R., p. 1059.

    d) Morris, D. G., p. 1060.
    e) Bargon, J., p. 1061.
    f) Closs, G. L., p. 1064.
    g) Buchachenko, A. L., p. 1065.
    h) Atlanti, P., Biellmann, J., Briere, R., Lemaire, H., Rassat, A., p. 1066.
    i) Roth, H. D., p. 1068.
    j) Lippmaa, E., Pehk, T., Saluvere, T., p. 1070.
    k) Hollaender, J., Neumann, W. P., p. 1072.
    l) Blank, B., Fischer, H., p. 1075.
    m) Rieker, A., p. 1078.
    n) Dominh, T., p. 1080.
    o) Den Hollander, J. A., p. 1083.
    p) Ward, H. R., p. 1085.
    q) Lawler, R. G., Evans, G. T., p. 1087.
103. Tomkiewicz, M., Cocivera, M.: Chem. Phys. Letters **8**, 595 (1971).
104. Adrian, F. J.: Chem. Phys. Letters **10**, 70 (1971).
105. Kessenikh, A. V., Rykov, S. V., Yankelevich, A. Z.: Chem. Phys. Letters **9**, 347 (1971).
106. Ignatenko, A. V., Kessenikh, A. V.: Org. Mag. Resonance **3**, 797 (1971).
107. Kobrina, L. S., Vlasova, L. V., Mamatyuk, V. I.: Izv. Sib. Otd. Akad. Nauk SSSR, Ser. Khim. Nauk. **1971**, 92.
108. Bodewitz, H. W. H. J., Blomberg, C., Bickelhaupt, F.: Tetrahedron Letters **1972**, 281.
109. Muller, K.: Chem. Commun. **1972**, 45.
110. Fahrenholtz, S. R., Trozzolo, A. M.: J. Am. Chem. Soc. **94**, 282 (1972).
111. a) Cooper, R. A., Lawler, R. G., Ward, H. R.: J. Am. Chem. Soc. **94**, 545 (1972).
     b) Cooper, R. A., Lawler, R. G., Ward, H. R.: J. Am. Chem. Soc. **94**, 552 (1972).
112. Ward, H. R.: Acc. Chem. Res. **5**, 18 (1972).
113. Lawler, R. G.: Acc. Chem. Res. **5**, 25 (1972).
114. Muller, K., Closs, G. L.: J. Am. Chem. Soc. **94**, 1002 (1972).
115. Roth, H. D., Lamola, A. A.: J. Am. Chem. Soc. **94**, 1013 (1972).
116. Walling, C., Lepley, A. R.: J. Am. Chem. Soc. **94**, 2007 (1972).
117. Morris, J. I., Morrison, R. C., Smith, D. W., Garst, J. F.: J. Am. Chem. Soc. **94**, 2406 (1972).
118. Bethell, D., Brinkman, M. R., Hayes, J.: Chem. Commun. **1972**, 475.
119. Porter, N. A., Marnett, L. J., Lochmuller, C. H., Closs, G. L., Shobataki, M.: J. Am. Chem. Soc. **94**, 3664 (1972).
120. Mitchell, T. N.: Tetrahedron Letters **1972**, 2281.
121. Berger, S., Hauff, S., Niederer, P., Rieker, A.: Tetrahedron Letters **1972**, 2581.
122. Tomkiewicz, M., Groen, A., Cocivera, M.: Chem. Phys. Letters **10**, 39 (1971); – J. Chem. Phys. **56**, 5850 (1972).
123. Deutch, J. M.: J. Chem. Phys. **56**, 6076 (1972).
124. Ankers, W. B., Brown, C., Hudson, R. F., Lawson, A. J.: Chem. Commun. **1972**, 935.
125. Kaplan, M. L., Roth, H. R.: Chem. Commun. **1972**, 970.
126. Freemann, R., Hill, H. D. W., Kaptein, R.: J. Mag. Res. **7**, 327 (1972).
127. Schulman, E. M., Bertrand, R. D., Grant, D. M., Lepley, A. R., Walling, C.: J. Am. Chem. Soc. **94**, 5972 (1972).
128. a) Kaptein, R.: J. Am. Chem. Soc. **94**, 6251 (1972).
     b) Kaptein, R.: J. Am. Chem. Soc. **94**, 6262 (1972).
129. Kaptein, R., Den Hollander, J. A.: J. Am. Chem. Soc. **94**, 6269 (1972).
130. Kaptein, R., Brokken-Zijp, J., De Kanter, F. J. J.: J. Am. Chem. Soc. **94**, 6280 (1972).
131. Schaffner, K., Wolf, H., Rosenfeld, S. M., Lawler, R. G., Ward, H. R.: J. Am. Chem. Soc. **94**, 6553 (1972).
132. Cocivera, M., Tomkiewicz, M., Groen, A.: J. Am. Chem. Soc. **94**, 6598 (1972).
133. Buchachenko, A. L., Markarian, Sh. A.: Int. J. Chem. Kinet. **4**, 513 (1972).
134. Richard, C.: Second Thesis of Doctorat ès Sciences Physiques, Nancy (1972).
135. Pine, S. H.: J. Chem. Ed. **49**, 664 (1972).
136. a) Gerhart, F., Ostermann, G.: Tetrahedron Letters **1969**, 4705.
     b) Gerhart, F.: Tetrahedron Letters **1969**, 5061.

137. IWAMURA, H., IWAMURA, M., SATO, S., KUSHIDA, K.: Bull. Chem. Soc. Japan **44**, 876 (1971).
138. HEESING, A., KAISER, B. U.: Tetrahedron Letters **1970**, 2845.
139. OOSTERHOFF, L. J.: Koninkl. Nederl. Akad. Wetenschap. (Amst.) **80**, 10 (1971).
140. ROTH, H. D.: J. Am. Chem. Soc. **93**, 1527 and 4935 (1971); **94**, 1400 and 1761 (1972).
141. MARUYAMA, K., SHINDO, H., MARUYAMA, T.: Bull. Chem. Soc. Japan **44**, 585 (1971).
142. MARUYAMA, K., OTSUKI, T., SHINDO, H., MARUYAMA, T.: Bull. Chem. Soc. Japan **44**, 2000 (1971).
143. BUCHACHENKO, A. L., ZHIDOMIROV, F. M.: Usp. Khim. **40**, 1729 (1971); – Russ. Chem. Rev. **40**, 801 (1971).
144. TOMKIEWICZ, M., KLEIN, M. P.: Rev. Sci. Inst. **43**, 1206 (1972).
145. NELSEN, S. F., METZLER, R. B., IWAMURA, M.: J. Am. Chem. Soc. **91**, 5103 (1969).
146. IWAMURA, H., IWAMURA, M., NISHIDA, T., SATO, S.: J. Am. Chem. Soc. **92**, 7474 (1970).
147. LEPLEY, A. R., BECKER, R. H., GIUMANINI, A. G.: J. Org. Chem. **36**, 1222 (1971).
148. SADLER, I. H.: Ann. Reports (B) **68**, 18 (1971).
149. LEHNIG, M., FISCHER, H.: Z. Naturforsch. **27a**, 1300 (1972).
150. TOMKIEWICZ, M., MCALPINE, R. D., COCIVERA, M.: Can. J. Chem. **50**, 3849 (1972).
151. ATLANI, P., BIELLMANN, J. F., BRIERE, R., RASSAT, A.: Tetrahedron **28**, 5805 (1972).
152. CLOSS, G. L., DOUBLEDAY, C. E.: J. Am. Chem. Soc. **94**, 9248 (1972).
153. ROSENFELD, S. M., LAWLER, R. G., WARD, H. R.: J. Am. Chem. Soc. **94**, 9255 (1972).
154. ATLANI, P., BIELLMANN, J. F., BRIERE, R., LEMAIRE, H., RASSAT, A.: Tetrahedron **28**, 2827 (1972).
155. a) BETHELL, D., BRINKMAN, M. R., HAYES, J.: Chem. Commun. **1972**, 1323.
     b) BETHELL, D., BRINKMAN, M. R., HAYES, J.: Chem. Commun. **1972**, 1324.
156. ATKINS, P. W., GURD, R. C.: Chem. Phys. Letters **16**, 265 (1972).
157. ATKINS, P. W., GURD, R. C., MOORE, E. A.: Chem. Phys. Letters **16**, 270 (1972).
158. BAKKER, B. H., SCHILDER, G. J. A., REINTS BOK, TH., STEINBERG, H., DE BOER, TH.: Tetrahedron **29**, 93 (1973).
159. ROTH, H. D., KAPLAN, M. L.: J. Am. Chem. Soc. **95**, 262 (1973).
160. LOKEN, H. Y., LAWLER, R. G., WARD, H. R.: J. Org. Chem. **38**, 106 (1973).
161. BLANK, B., FISCHER, H.: Helv. Chim. Acta **56**, 506 (1973).
162. BARGON, J.: J. Am. Chem. Soc. **95**, 941 (1973).
163. ROSENFELD, S. M., LAWLER, R. G., WARD, H. R.: J. Am. Chem. Soc. **95**, 946 (1973).
164. SPENCER, T. S., DONNEL, C. M. O.: J. Chem. Ed. **50**, 152 (1973).
165. ALLEN, R. B., LAWLER, R. G., WARD, H. R.: J. Am. Chem. Soc. **95**, 1692 (1973).
166. BRINKMAN, M. R., BETHELL, D., HAYES, J.: Tetrahedron Letters **1973**, 989.
167. MARUYAMA, K., OTSUKI, T., TAKUWA, A.: Chemistry Letters **1972**, 131.
168. DOERFFEL, K., HÖBOLD, W., HORN, R.: J. Prakt. Chem. **313**, 991 (1971).
169. TOMKIEWICZ, M., KLEIN, M. P.: Proc. Nat. Acad. Sci. US. **70**, 143 (1973).
170. a) FESSENDEN, R. W., SCHULER, R. H.: J. Chem. Phys. **39**, 2147 (1963).
     b) SMALLER, B., REMKO, J. R., AVERY, E. C.: J. Chem. Phys. **48**, 5174 (1968).
     c) PAUL, H., FISCHER, H.: Z. Naturforsch. **25a**, 443 (1970).
     d) GLARUM, S. H., MARSHALL, J. H.: J. Chem. Phys. **52**, 5555 (1970).
     e) LIVINGSTON, R., ZELDES, H.: J. Chem. Phys. **53**, 1406 (1970).
     f) ATKINS, P. W., BUCHANAN, I. C., GURD, R. C., MCLAUCHLAN, K. A., SIMPSON, A. F.: Chem. Commun. **1970**, 513.
     g) ADRIAN, F. J.: J. Chem. Phys. **54**, 3918 (1971).
     h) NETA, P., FESSENDEN, R. W., SCHULER, R. H.: J. Phys. Chem. **75**, 1654 (1971).
     i) SMALLER, B., AVERY, E. C., REMKO, J. R.: J. Chem. Phys. **55**, 2414 (1971).
     j) EIBEN, K., FESSENDEN, R. W.: J. Phys. Chem. **75**, 1186 (1971).
     k) ATKINS, P. W., GURD, R. C., MCLAUCHLAN, K. A., SIMPSON, A. F.: Chem. Phys. Letters **8**, 55 (1971).
     l) PEDERSEN, J. B., FREED, J. H.: J. Chem. Phys. **57**, 1004 (1972).
     m) WONG, S. K., WAN, J. K. S.: J. Am. Chem. Soc. **94**, 7197 (1972).
     n) HUTCHINSON, D. A., WONG, S. K., COLPA, J. P., WAN, J. K. S.: J. Chem. Phys. **57**, 3308 (1972).
     o) ADRIAN, F. J.: J. Chem. Phys. **57**, 5107 (1972).

p) WONG,S.K., HUTCHINSON,D.A., WAN,J.K.S.: J. Am. Chem. Soc. **95**, 622 (1973).

q) ATKINS,P.W., MCLAUCHLAN, K.A., PERCIVAL,P.W.: Chem. Commun. **1973**, 121.

r) WONG,S.K., HUTCHINSON,D.A., WAN,J.K.S.: J. Chem. Phys. **58**, 985 (1973).

s) PEDERSEN,J.B., FREED,J.H.: J. Chem. Phys. **58**, 2746 (1973).

t) ATKINS,P.W.: Chem. Phys. Letters **18**, 290 (1973).

u) ATKINS,P.W.: Org. Mag. Res. **5**, 239 (1973).

v) SHUVALOV,V.F., STUNZHAS,P.A., MORAVSKII,A.P.: Org. Mag. Res. **5**, 347 (1973).

w) PEDERSEN,J.B., FREED,J.H.: J. Chem. Phys. **59**, 2869 (1973).

x) ATKINS,P.W., MOORE,E.A.: Mol. Phys. **25**, 825 (1973).

y) WONG,S.K., WAN,J.K.S.: J. Chem. Phys. **59**, 3859 (1973).

z) ATKINS,P.W., DOBBS,A.J., MCLAUCHLAN,K.A.: Chem. Phys. Letters **22**, 209 (1973).

171. EVANS,G.T., FLEMMING III.P.D., LAWLER,R.G.: J. Chem. Phys. **58**, 2071 (1973).

172. CLOSS,G.L., DOUBLEDAY,C.E.: J. Am. Chem. Soc. **95**, 2735 (1973).

173. TOMKIEWICZ,M., KLEIN,M.P.: J. Am. Chem. Soc. **95**, 3132 (1973).

174. HARGIS,J.H., SHEVLIN,P.B.: Chem. Commun. **1973**, 179.

175. ROTH,H.D.: Mol. Photochemistry **5**, 91 (1973).

176. Chemically Induced Magnetic Polarization. Eds.: CLOSS,G.L., LEPLEY,A.R. New York: Wiley 1973.
　　a) GLARUM,S.H., p. 1.
　　b) ATKINS,P.W., MCLAUCHLAN,K.A., p. 41.
　　c) CLOSS,G.L., p. 95.
　　d) KAPTEIN,R., p. 137.
　　e) FISCHER,H., p. 197.
　　f) GARST,J.F., p. 223.
　　g) WARD,H.R., LAWLER,R.G., COOPER,R.A., p. 281.
　　h) LEPLEY,A.R., p. 323.

177. IWAMURA,H., IWAMURA,M., IMANARI,M., TAKEUCHI,M.: Tetrahedron Letters **1973**, 2325.

178. IGNATENKO,A.V., KESSENIKH,A.V., GLUKHOVTSEV,V.G., NADTOCHII,M.A.: Org. Mag. Res. **5**, 219 (1973).

179. KESSENIKH,A.V., PETROVSKII,P.V., RYKOV,S.V.: Org. Mag. Res. **5**, 227 (1973).

180. BUCHACHENKO,A.L., MARKARYAN,SH.A.: Org. Mag. Res. **5**, 247 (1973).

181. ADAM,W., ARCE DE SANABIA,J., FISCHER,H.: J. Org. Chem. **38**, 2571 (1973).

182. NIKIFOROV,G.A., MARKARYAN,SH.A., PLEKHANOVA,L.G., SVIRIDOV,B.D., RYKOV, S.V., ERSHOV,V.V., BUCHACHENKO,A.L., PEHK,T., SALUVERE,T., LIPPMAA,E.: Org. Mag. Res. **5**, 339 (1973).

183. RUDAKOV,E.S., MASTIKHIN,V.M., POPOV,S.G., RUDAKOVA,R.I.: Org. Mag. Res. **5**, 343 (1973).

184. ALLEN,R.B., LAWLER,R.G., WARD,H.R.: Tetrahedron Letters **1973**, 3303.

185. BROWN,C., HUDSON,R.F., LAWSON,A.J.: J. Am. Chem. Soc. **95**, 6500 (1973).

186. LIPPMAA,E., SALUVERE,T., PEHK,T., OLIVSON,A.: Org. Mag. Res. **5**, 429 (1973).

187. BUBNOV,N.N., MEBVEDEV,B.YA., POLYAKOVA,L.A., BILEVICH,K.A., OKHLOBYSTIN, O.YU.: Org. Mag. Res. **5**, 437 (1973).

188. LIPPMAA,E., PEHK,T., SALUVERE,T., MÄGI,M.: Org. Mag. Res. **5**, 441 (1973).

189. GRAGEROV,I.P., LEVIT,A.F., KIPRIANOVA,L.A., BUCHACHENKO,A.L.: Org. Mag. Res. **5**, 445 (1973).

190. SAVIN,V.I., TEMYACHEV,I.D., KITAEV,YU.P.: Org. Mag. Res. **5**, 449 (1973).

191. OLLIS,W.B., SUTHERLAND,I.O., THEBTARANONTH,Y.: Chem. Commun. **1973**, 653.

192. OLLIS,W.B., SUTHERLAND,I.O., THEBTARANONTH,Y.: Chem. Commun. **1973**, 654.

193. CHEN,H.E.C., GROEN,A., COCIVERA,M.: Can. J. Chem. **51**, 3032 (1973).

194. MARUYAMA,K., OTSUKI,T., TAKUWA,A., ARAKAWA,S.: Bull. Chem. Soc. Japan **46**, 2470 (1973).

195. ADAM,W., FISCHER,H., HANSEN,H.J., HEIMGARTNER,H., SCHMID,H., WAESPE,H.R.: Angew. Chem. Int. Ed. (Engl.) **12**, 662 (1973).

196. DEUTCH,J.M.: J. Chem. Phys. **59**, 2762 (1973).

197. GONZENBACH, H. U., SCHAFFER, K., BLANK, B., FISCHER, H.: Helv. Chim. Acta **56**, 1741 (1973).
198. BRINKMAN, M. R., BETHELL, D., HAYES, J.: J. Chem. Phys. **59**, 3431 (1973).
199. LEVIN, YA. A., IL'YASOV, A. V., GOLDFARB, E. I., VORKUNOVA, E. I.: Org. Mag. Res. **5**, 487 (1973).
200. LEVIN, YA. A., IL'YASOV, A. V., GOLDFARB, E. I., VORKUNOVA, E. I.: Org. Mag. Res. **5**, 497 (1973).
201. POBEDIMSKY, D. G., KIRPICHNIKOV, P. A., SAMITOV, YU. YU., GOLDFARB, E. I.: Org. Mag. Res. **5**, 503 (1973).
202. SEIFERT, K. G., BARGON, J.: Angew. Chem. Int. Ed. Engl. **12**, 763 (1973).
203. CHEN, H. E., VAISH, S. P., COCIVERA, M.: J. Am. Chem. Soc. **95**, 7586 (1973).
204. CHEN, H. E., VAISH, S. P., COCIVERA, M.: Chem. Phys. Letters **22**, 576 (1973).
205. KESSENIKH, A. V., IGNATENKO, A. V., RYKOV, S. V.: Org. Mag. Res. **5**, 533 (1973).
206. KESSENIKH, A. V., IGNATENKO, A. V., RYKOV, S. V., SHTEINSHNEIDER, A. YA.: Org. Mag. Res. **5**, 537 (1973).
207. BELETSKAYA, I. P., RYKOV, S. V., BUCHACHENKO, A. L.: Org. Mag. Res. **5**, 595 (1973).
208. SAGDEEV, R. Z., MOLIN, YU. N., SALIKHOV, K. M., LESHINA, T. V., KAMHA, M. A., SHEIN, S. M.: Org. Mag. Res. **5**, 599 (1973).
209. SAGDEEV, R. Z., MOLIN, YU. N., SALIKHOV, K. M., LESHINA, T. V., KAMHA, M. A., SHEIN, S. M.: Org. Mag. Res. **5**, 603 (1973).
210. KAPTEIN, R., FREEMAN, R., HILL, H. D. W., BARGON, J.: Chem. Commun. **1973**, 953.
211. IWAMURA, H., IWAMURA, M., IMANARI, M., TAKEUCHI, M.: Bull. Chem. Soc. Japan **46**, 3486 (1973).
212. Papers presented at the International Symposium on Chemically Induced Polarization of Nuclei and Electrons, Tallin, Estonian SSR, August 13—16, 1972 and published in Org. Mag. Res. (1973) (see previous references).
213. Free Radicals, Vol. 1, Dynamics of Elementary Processes, Edited by KOCHI, J. K. New York (1973).
    a) WARD, H. R., p. 239.
    b) WILT, J. W., p. 333.
214. ATKINS, P. W., FRIMSTON, J. M., FRITH, P. G., GURD, R. C., MCLAUCHLAN, K. A.: J.C.S. Faraday Trans. II, **69**, 1542 (1973).
215. TSUJI, T., NISHIDA, S.: Chemistry Letters **1973**, 1335.
216. BARGON, J., SEIFERT, K. G.: J. Phys. Chem. **77**, 2877 (1973).
217. SCHÄUBLIN, S., HÖHENER, A., ERNST, R. R.: J. Mag. Res. (to be published).

Subject Index

NMR

**Basic Principles and Progress
Grundlagen und Fortschritte**

Editors:
P. Diehl, E. Fluck, R. Kosfeld

**Vol. 1: P. Diehl, C. L. Kethrapal:
NMR Studies of Molecules
Oriented in the Nematic Phase of
Liquid Crystals
R. G. Jones: The Use of Symmetry
in Nuclear Magnetic Resonance**
53 figs. V, 174 pages. 1969
Cloth DM 39,–; US $15.10
ISBN 3-540-04665-8

**Vol. 2: H. J. Keller: NMR-
Untersuchungen an Komplex-
verbindungen**
22 Abb. III, 88 Seiten. 1970
Geb. DM 32,–; US $12.40
ISBN 3-540-04980-0

**Vol. 3: O. Kanert, M. Mehring:
Static Quadrupole Effects in
Disordered Cubic Solids
F. Noack: Nuclear Magnetic
Relaxation Spectroscopy**
73 figs. V, 144 pages. 1971
Cloth DM 48,–; US $18.50
ISBN 3-540-05392-1

**Vol. 4: Natural and Synthetic
High Polymers**
Lectures presented at the Seventh
Colloquium on NMR Spectroscopy.
Held in the Institut
für Physikalische Chemie,
April 13-17, 1970,
Technische Hochschule Aachen
202 figs. X, 309 pages. 1971
Cloth DM 64,–; US $24.70
ISBN 3-540-05221-6

**Vol. 5: R. A. Hoffman, S. Forsén,
B. Gestblom: Analysis of NMR
Spectra**
A Guide for Chemists
63 figs. III, 165 pages. 1971
Cloth DM 48,–; US $18.50
ISBN 3-540-05427-8

**Vol. 6: P. Diehl, H. Kellerhals,
E. Lustig:
Computer Assistance in
the Analysis of High-Resolution
NMR Spectra**
11 figs. III, 96 pages. 1972
Cloth DM 48,–; US $18.50
ISBN 3-540-05532-0

**Vol. 7: C. W. Hilbers, C. MacLean:
NMR of Molecules Oriented in
Electric Fields
H. Pfeifer: Nuclear Magnetic
Resonance and Relaxation of
Molecules Adsorbed on Solids**
56 figs. V, 153 pages. 1972
Cloth DM 78,–; US $30.10
ISBN 3-540-05687-4

Prices are subject to change
without notice

**Springer-Verlag
Berlin Heidelberg New York**

München Johannesburg London
Madrid New Delhi Paris
Rio de Janeiro Sydney Tokyo
Utrecht Wien

Topics in Current Chemistry

Fortschritte der chemischen Forschung

Springer-Verlag Berlin · Heidelberg · New York